THE NUCLEAR ENTERPRISE

The Hoover Institution gratefully acknowledges the following individuals and foundations for their significant support of this publication and the conference on which this book is based:

THOMAS AND BARBARA STEPHENSON

THE WILLIAM AND FLORA HEWLETT FOUNDATION

PRESTON AND CAROLYN BUTCHER

JOHN AND ANN DOERR

THE KORET FOUNDATION

THE MARY JO AND DICK KOVACEVICH FAMILY FOUNDATION

WILLIAM AND SUSAN OBERNDORF

PAUL AND SANDRA OTELLINI

THE THOMAS AND STACEY SIEBEL FOUNDATION

THE NUCLEAR ENTERPRISE

High-Consequence Accidents:
How to Enhance Safety and Minimize
Risks in Nuclear Weapons and Reactors

Edited by

George P. Shultz *and* Sidney D. Drell

HOOVER INSTITUTION PRESS
STANFORD UNIVERSITY STANFORD, CALIFORNIA

The Hoover Institution on War, Revolution and Peace, founded at Stanford University in 1919 by Herbert Hoover, who went on to become the thirty-first president of the United States, is an interdisciplinary research center for advanced study on domestic and international affairs. The views expressed in its publications are entirely those of the authors and do not necessarily reflect the views of the staff, officers, or Board of Overseers of the Hoover Institution.

www.hoover.org

Hoover Institution Press Publication No. 626

Hoover Institution at Leland Stanford Junior University,
Stanford, California 94305-6010

First printing 2012
18 17 16 15 14 13 12 7 6 5 4 3 2 1

Manufactured in the United States of America

The paper used in this publication meets the minimum Requirements of the American National Standard for Information Sciences—Permanence of Paper for Printed Library Materials, ANSI/NISO Z39.48-1992. ♾

Cataloging-in-Publication Data is available from the Library of Congress.
ISBN: 978-0-8179-1524-7 (cloth. : alk. paper)
ISBN: 978-0-8179-1525-4 (pbk. : alk. paper)
ISBN: 978-0-8179-1526-1 (e-book)

Contents

Abbreviations

AEC	Atomic Energy Commission
ALWR	Advanced Light Water Reactor
BDBEs	beyond-design basis events
BRC	Blue Ribbon Commission on America's Nuclear Future
BWR	boiling water reactor
DBEs	design basis events
DoD	Department of Defense
DOE	Department of Energy
ENDS	enhanced nuclear detonation safety
EPRI	Electric Power Research Institute
FERC	Federal Energy Regulatory Commission
FOAK	first-of-a-kind
GHG	greenhouse gas emissions
GWe	gigawatts of electricity
HE	high explosive
IAEA	International Atomic Energy Agency
IEA	International Energy Agency
IHE	insensitive high explosive
INES	International Nuclear and Radiological Event Scale
INPO	Institute of Nuclear Power Operations
INSAG	International Nuclear Safety Advisory Group
LLTF	Lessons Learned Task Force

LOCA	loss of coolant accident
LTBT	Limited Test Ban Treaty
LWR	light water reactor
MWe	megawatts of electricity
NEA	Nuclear Energy Agency
NGO	non-governmental organization
NOAK	nth-of-a-kind
NPT	Non-Proliferation Treaty
NRC	Nuclear Regulatory Commission
NSG	Nuclear Suppliers Group
OECD	Organisation for Economic Co-operation and Development
PRA	probabilistic risk assessment
PWR	pressurized water reactor
R&D	research and development
RBMK	Reaktor Bolshoy Moshchnosti Kanalniy
RD&D	research, development & demonstration
RDT&E	research, development, test, and evaluation
SMR	small modular reactor
TEPCO	Tokyo Electric Power Company
TMI	Three Mile Island
UNSCEAR	United Nations Scientific Committee on the Effects of Atomic Radiation
WANO	World Association of Nuclear Operators

List of Figures and Tables

Figures

Tables

Preface

Nuclear energy can provide great benefits to society; electric power and nuclear medical applications are clear examples. In the much different form of nuclear weapons, however, it can cause death and destruction on an unparalleled scale. These activities form the nuclear enterprise that has been at the center of mankind's dreams and nightmares since it burst upon the world's consciousness on August 6, 1945.

For many years, there has been a general reluctance to use the word "nuclear" when discussing weapons or civilian applications of nuclear energy. Note how the word is buried, such as in the description of magnetic resonance imaging (MRI). This valuable medical diagnostic tool is, in fact, an application of *nuclear* magnetic resonance technology. It is imperative that all uses of nuclear energy be executed with great care to prevent unintended consequences such as the events following the recent tragedy in Japan. Almost 20,000 people were killed or reported missing in the mammoth earthquake and subsequent tsunami, and many more had to be evacuated due to the escape of radioactive material from crippled nuclear power reactors at Fukushima. The disaster has prompted an international reappraisal of the trade-offs between the costs, benefits, and risks of the nuclear power enterprise.

Nuclear arsenals around the world today present a potentially analogous situation. We are collectively at risk of some precipitating event—be it equipment or human failure, miscommunication or miscalculation, or a deliberate deadly act by suicidal terrorists—producing catastrophic consequences with nuclear devastation far beyond the scale inflicted by the natural disaster in Japan.

The challenge before us is how to deal with the catastrophic risk of the nuclear enterprise in a way that preserves its positive elements and makes economic sense.

- What can and should be done to improve operations and public understanding of the risks and consequences of major incidents?
- How can informed scientists, economists, and journalists interact more effectively in understanding and reporting to the public on the most important issues affecting risks, consequences, and costs?

In addressing this topic of activities that present special risks, which are rare but have the potential for catastrophic consequences, we are reminded of the book *Risk & Other Four-Letter Words*, written by the late Walter B. Wriston in 1986. He was careful to distinguish "risk" from "recklessness," but concluded with these paragraphs:

> Let those who seek a perpetual safe harbor continue to do so. Let them renounce risk for themselves, if they choose. What no one has a right to do is renounce it for all the rest of us, or to pursue the chimerical goal of a risk-free society for some by eliminating the rewards of risk for everyone.
>
> The society that promises no risks, and whose leaders use the word "risk" only as a pejorative, may be able to protect life, but there will be no liberty, and very little pursuit of happiness. You will look in vain in the Federalist Papers, the Declaration of Independence, or the Constitution for promises of a safe, easy, risk-free life. Indeed, when Woodrow Wilson called for a world safe for democracy, it was left to Gilbert Chesterton to put that sentiment in perspective. "Impossible," he said. "Democracy is a dangerous trade."

The risks and rewards of the nuclear enterprise were the subject of a conference held at Stanford University's Hoover Institution on October 3–4, 2011. Papers for this conference were prepared by specialists on various aspects of this challenging topic, including technical safety, management operations, regulatory measures, and the importance of accurate communication by the media. This book contains these papers, edited by the authors in response to discussions at the conference.

It is our judgment that the global dangers posed by the nuclear enterprise are growing. It is our hope that the findings of the conference will contribute to discussion and then action to better control and contain those dangers.

Acknowledgments

This book on the nuclear enterprise is the result of a conference held at the Hoover Institution at Stanford University on October 3–4, 2011. The editors thank the authors for writing and submitting excellent papers for that conference and for editing them in response to discussions at the conference for publication in this book. Many individuals at the Hoover Institution and Hoover Institution Press made valuable contributions to the success of the conference and preparation of this book. We want to thank Summer Tokash for her excellent oversight of every detail of this conference—before, during, and after. She was the essential initiator, coordinator, and loop-closer who made it all possible. In particular, we wish to thank Barbara Egbert for the exceptional work she did in editing the manuscripts and bringing them into their final form and Barbara Arellano for her patience and valuable contributions as book production manager.

Introduction

SIDNEY D. DRELL, GEORGE P. SHULTZ, AND
STEVEN P. ANDREASEN

Policy Overview

We live in dangerous times for many reasons. Prominent among them is the existence of a global nuclear enterprise made up of weapons that can cause damage of unimaginable proportions and power plants at which accidents can have severe, essentially unpredictable consequences for human life. For all of its utility and promise, the nuclear enterprise is unique in the enormity of the vast quantities of destructive energy that can be released through blast, heat, and radioactivity.

We addressed just this subject in a conference in October 2011 at Stanford University's Hoover Institution. The complete set of papers prepared for the conference is reproduced in this book. The conference included experts on weapons, on power plants, on regulatory experience, and on the development of public perceptions and the ways in which these perceptions influence policy. The reassuring outcome of the conference was a general sense that the U.S. nuclear enterprise currently meets very high standards in its commitment to safety and security.

That has not always been the case in all aspects of the nuclear enterprise. And the unsettling outcome of the conference was that it will not be the case globally unless governments, international organizations, industry, and media recognize and address the nuclear challenges and mounting risks posed by a rapidly changing world.

The acceptance of the nuclear enterprise is now being challenged by concerns about the questionable safety and security of programs primarily in countries relatively new to the nuclear enterprise, and the potential loss of control to terrorist or criminal gangs of fissile material that exists in such abundance around the world. In a number of countries, confidence in nuclear energy production was severely shaken in the spring of 2011 by the Fukushima nuclear reactor plant disaster. And in the military sphere, the doctrine of deterrence that remains primarily dependent on nuclear weapons is seen in decline due to the importance of non-state actors such as al Qaeda and terrorist affiliates that seek destruction for destruction's sake. We have two nuclear tigers by the tail.

When risks and consequences are unknown, undervalued, or ignored, our nation and the world are dangerously vulnerable. Nowhere is this risk-consequence equation more relevant than with respect to the nucleus of the atom.

The nuclear enterprise was introduced to the world by the shock of the devastation produced by two atomic bombs hitting Hiroshima and Nagasaki. Modern nuclear weapons are far more powerful than those early bombs, which presented their own hazards. Early research depended on a program of atmospheric testing of nuclear weapons. In the early years following World War II, the impact and the amount of radioactive fallout in the atmosphere generated by above-ground nuclear explosions was not fully appreciated. During those years, the United States and also the Soviet Union conducted several hundred tests in the atmosphere that created fallout. The recent Stanford conference focused on a regulatory weak point from that time that exists in many places today, as the Fukushima disaster clearly indicates. The U.S. Atomic Energy Commission (AEC) was initially assigned conflicting responsibilities: to create an arsenal of nuclear weapons for the United States to confront a growing nuclear-armed Soviet threat; and, at the same time, to ensure public safety from the effects of radioactive fallout. The AEC was faced with the same conundrum with regard to civilian nuclear power generation. It was charged with promoting civilian nuclear power and simultaneously protecting the public.

Progress came in 1963 with the negotiation and signing of the Limited Test Ban Treaty (LTBT) banning all nuclear explosive testing in the atmosphere (initially by the United States, the Soviet Union, and the United Kingdom). With the successful safety record of the U.S. nuclear weapons program, domestic anxiety about nuclear weapons receded somewhat. Meanwhile, public attitudes toward nuclear weapons reflected recognition of their key role in establishing a more stable nuclear deterrent posture in the confrontation with the Soviet Union.

The positive record on safety of the nuclear weapons enterprise in the United States—there have been accidents involving nuclear weapons, but none that led to the release of nuclear energy—was the result of a strong effort and continuing commitment to include safety as a primary criterion in new weapons designs, as well as careful production, handling, and deployment procedures. The key to the health of today's nuclear weapons enterprise is confidence in the safety of its operations and in the protection of special nuclear materials against theft. One can imagine how different the situation would be today if there had been a recognized theft of material sufficient for a bomb, or if one of the two four-megaton bombs dropped from a disabled B-52 Strategic Air Command bomber overflying Goldsboro, North Carolina, in 1961 had detonated. In that event, just one switch in the arming sequence of one of the bombs, by remaining in its "off position" while the aircraft was disintegrating, was all that prevented a full-yield nuclear explosion. A close call indeed!

In the twenty-six years since Chernobyl, the nuclear power industry has strengthened its safety practices. Over the past decade, growing concerns about global warming and energy independence have actually strengthened support for nuclear energy in the United States and many nations around the world. Yet despite these trends, the civil nuclear enterprise remains fragile. Following Fukushima, opinion polls gave stark evidence of the public's deep fears of the invisible force of nuclear radiation, shown by public opposition to the construction of new nuclear power plants in close proximity. It is not simply a matter of getting better information to the public but of actually educating the public about the true nature of nuclear radiation and its risks. Of course, the immediate task of the nuclear power component of the enterprise is to strive for the best possible safety record with one overriding objective: no more Fukushimas.

Another issue that must be resolved involves the continued effectiveness of a policy of deterrence that remains primarily dependent upon nuclear weapons, and the hazards these weapons pose due to the spread of nuclear technology and material. There is growing apprehension about the determination of terrorists to get their hands on weapons or, for that matter, on the special nuclear material—plutonium and highly enriched uranium—that fuels them in the most challenging step toward developing a weapon.

The global effects of a regional war between nuclear-armed adversaries such as India and Pakistan would also wield an enormous impact, potentially involving radioactive fallout at large distances caused by a limited number of nuclear explosions.

This is true as well for nuclear radiation from a reactor explosion—fallout at large distances would have a serious societal impact on the nuclear enterprise. There is little understanding of the reality and potential danger of consequences if such an event were to occur halfway around the world. An effort should be made to prepare the public by providing information on how to respond to such an event.

An active nuclear diplomacy has grown out of the Cold War efforts to regulate testing and reduce superpower nuclear arsenals. There is now a welcome focus on rolling back nuclear weapons proliferation. Additional important measures include the Nunn-Lugar program, started in 1991 to reduce the nuclear arsenal of the former Soviet Union. These have led to greater investment by the U.S. and other governments in better security for nuclear weapons and material globally, including billions of dollars through the G8 Global Partnership Against the Spread of Weapons and Materials of Mass Destruction. The commitment to improving security of all dangerous nuclear material on the globe within four years was made by forty-seven world leaders who met with President Obama in Washington, D.C., in April 2010; this commitment was reconfirmed in March 2012 at the Nuclear Security Summit in Seoul, South Korea. Many specific commitments made in 2010 relating to the removal of nuclear materials and conversion of nuclear research reactors from highly enriched uranium to low-enriched uranium fuel have already been accomplished, along with increasing levels of voluntary commitments from a diverse set of states, improving prospects for achieving the four-year goal.

The nuclear enterprise faces new and increasingly difficult challenges. Successful leadership in national security policy will require a continuous, diligent, and multinational assessment of these newly emerging risks and consequences.

The Stanford conference examined the risks and potentially deadly consequences associated with nuclear weapons and nuclear power, and identified three guiding principles for efforts to reduce those risks globally:

First, the calculations used to assess nuclear risks in both the military and the civil sectors are fallible. Accurately analyzing events where we have little data, identifying every variable associated with risk, and the possibility of a single variable that goes dangerously wrong are all factors that complicate risk calculations. Governments, industry, and concerned citizens must constantly re-examine the assumptions on which safety and security measures, emergency preparations, and nuclear energy production are based. When dealing with very low-probability and high-consequence operations, we typically have little data as a basis for making quantitative analyses. It is therefore difficult to assess the risk of a nuclear accident and what would contribute to it, and to identify effective steps to reduce that risk.

In this context, it is possible that a single variable could exceed expectations, go dangerously wrong, and simply overwhelm safety systems and the risk assessments on which those systems were built. This is what happened in 2011 when an earthquake, followed by a tsunami—both of which exceeded expectations based on history—overwhelmed the Fukushima complex, breaching a number of safeguards that had been built into the plant and triggering reactor core meltdowns and radiation leaks. This in turn exposed the human factor, which is hard to assess and can dramatically change the risk equation. Cultural habits and regulatory inadequacy inhibited rapid decision-making and crisis management in the Fukushima disaster. A more nefarious example of the human factor would be a determined nuclear terrorist attack specifically targeting either the military or civilian component of the nuclear enterprise.

Second, risks associated with nuclear weapons and nuclear power will likely grow substantially as nuclear weapons and civilian nuclear energy production technology spread in unstable regions of the world where the potential for conflict is high. States that are new to the nuclear enterprise may not have effective nuclear safeguards to secure nuclear weapons and materials—including a developed fabric of early warning systems and nuclear confidence-building measures that could increase warning and decision time for leaders in a crisis—or the capability to safely manage and regulate the construction and operation of new civilian reactors. Hence there is a growing risk of accidents, mistakes, or miscalculations involving nuclear weapons, and of regional wars or nuclear terrorism. The consequences would be horrific: a Hiroshima-size nuclear bomb detonated in a major city could kill a half-million people and result in $1 trillion in direct economic damage.

On the civil side of the nuclear ledger, the sobering paradox identified at the Stanford conference is this: while an accident would be considerably less devastating than the detonation of a nuclear weapon, the risk of an accident occurring is probably higher. Currently, 1.4 billion people live without electricity, and by 2030 the global demand for energy is projected to rise by about 25 percent. With the added need to minimize carbon emissions, nuclear power reactors will become increasingly attractive alternative sources for electric power, especially for developing nations. These countries, in turn, will need to meet the challenge of developing appropriate governmental institutions and the infrastructure, expertise, and experience to support nuclear power efforts with a suitably high standard of safety. As the world witnessed in Fukushima, a nuclear power plant accident can lead to the spread of dangerous radiation, massive civil dislocations, and billions of dollars in cleanup costs. Such an event can also fuel widespread public skepticism about nuclear institutions and technology.

Some developed nations—notably Germany—have interpreted the Fukushima accident as proof that they should abandon nuclear power altogether, primarily by prolonging the life of existing nuclear reactors while phasing out nuclear-produced electricity and developing alternative energy sources.

Third, we need to understand that no nation is immune from risks involving nuclear weapons and nuclear power within their borders. There were 32 so-called "Broken Arrow" accidents involving U.S. nuclear weapons between 1950 and 1980, mostly involving U.S. Strategic Air Command bombers and earlier bomb designs not yet incorporating modern nuclear detonation safety designs. The U.S. no longer maintains a nuclear-armed in-air strategic bomber force and the record of incidents is greatly reduced. In several cases, accidents such as the North Carolina bomber incident came dangerously close to triggering catastrophes, with disaster averted simply by luck.

The United States has had an admirable safety record in the area of civil nuclear power since the 1979 Three Mile Island accident in Pennsylvania, yet safety concerns persist. One of the critical assumptions in the design of the Fukushima reactor complex was that, if electrical power were lost at the plant and its back-up generators, it could be restored within a few hours. The combined one-two punch of the earthquake and tsunami, however, made such repair impossible. In the United States today, some nuclear power reactors are designed with a comparably short window for restoring power. After Fukushima, this is an issue that deserves action—especially in light of our own Hurricane Katrina experience, which rendered many affected areas inaccessible for days in 2005, and the August 2011 East Coast earthquake that shook the North Anna nuclear power plant in Mineral, Virginia, beyond expectations based on previous geological activity.

To reduce these nuclear risks, the conference arrived at four related recommendations that should be adopted by the nuclear enterprise, both military and civilian, in the United States and abroad.

First, the reduction of nuclear risks requires every level of the nuclear enterprise and related military and civilian organizations to embrace the importance of safety and security as an overarching operating rule. This is not as easy as it sounds. To a war fighter, more safety and control can mean less reliability and availability and greater costs. For a company or utility involved in the construction or operation of a nuclear power plant, more safety and security can mean greater regulation and higher costs.

But the absence of a culture of safety and security, in which priorities and meaningful standards are set and rigorous discipline and accountability are enforced, is perhaps the most reliable indicator of an impending disaster. In August 2007, after a B-52 bomber loaded with six nuclear-tipped cruise missiles flew from North Dakota to Louisiana without anyone realizing there were live weapons on board, then Secretary of Defense Robert Gates fired both the military and civilian heads of the U.S. Air Force. His action was an example of setting the right priorities and enforcing accountability, but the reality of the incident shows that greater incorporation of a safety and security culture is needed.

Second, independent regulation of the nuclear enterprise is crucial to setting and enforcing the safety and security rule. In the United States today, the nuclear regulatory system—in particular, the Nuclear Regulatory Commission (NRC)—is credited with setting a uniquely high standard for independent regulation of the civil nuclear power sector. This is one of the keys to a successful and safe nuclear program. Effective regulation is even more crucial when there are strong incentives to keep operating costs down and keep an aging nuclear reactor fleet in operation, a combination that could create conditions for a catastrophic nuclear power plant failure. Careful attention is required to protect the NRC from regulatory capture by vested interests in government and industry, the latter of which funds a high percentage of the NRC'S budget.

Strong, independent regulatory agencies are not the norm in many countries. The independent watchdog organization advising the Japanese government was working with Japanese utilities to influence public opinion in favor of nuclear power. Strengthening the International Atomic Energy Agency (IAEA) so that it can play a greater role in civil nuclear safety and security would also help reduce risks, and will require substantially greater authorities to address both safety and security, and most importantly resources for an agency whose budget is only 333 million Euros, with only 1/10th of that total going to nuclear safety and security. In addition, exporting "best practices" of the NRC—that is, lessons of nuclear regulation, oversight, and safety learned over many decades—to other countries would pay a huge safety dividend.

Third, independent peer review should be incorporated into all aspects of the nuclear enterprise. On the weapons side, independent experts in the United States—both within and outside the organization—are relied on to review, or "red team," each other, rigorously challenge and discuss weapons and systems safety, and communicate these points up and down the line. The Institute of Nuclear Power Operations (INPO) provides strong peer review and oversight of the civil nuclear sector in the United States. Its global counterpart, the World Association of Nuclear Operators (WANO), should give a higher priority to further strengthening its safety operations, in particular its peer review process, learning from the experiences of the United States and other nations. Strong outside peer review—combined with an enhanced capacity to arrange fines based on incidents occurring in far distant countries—would help states entering into the world of high-consequence operations to develop a culture and standard needed to achieve a high safety record.

Beyond these recommendations, the military and civilian nuclear communities can and should learn from each other. A periodic dialogue structured around assessing and reducing the risks surrounding the nuclear enterprise would be valuable, both in the United States and abroad, and could be organized by governments or academia (as was done in the conference at Stanford). An analysis of the probabilities of undesired events and ways to minimize them, including lessons learned from accidents such as Fukushima as well as "close call" incidents, should be put on the front burner along with consequence management—that is, what to do if a nuclear incident were to occur.

An informed public is also an essential element in responding to a nuclear crisis. Greater public awareness and understanding of nuclear risks and consequences can lead to greater public preparation to handle post-disaster challenges.

Fourth, progress on all aspects of nuclear threat reduction should be organized around a clear goal: a global effort to reduce reliance on nuclear weapons, prevent their spread into potentially dangerous hands, and ultimately end them as a threat to the world. A step-by-step process—along lines proposed by George Shultz, William Perry, Henry Kissinger, and Sam Nunn in a series of *Wall Street Journal* essays[1]—and demonstrated progress toward realizing the vision of a world free of nuclear weapons will build the kind of international trust and broad cooperation required to effectively address today's threats—and prevent tomorrow's catastrophe.

Our bottom line: Since the risks posed by the nuclear enterprise are so high, no reasonable effort should be spared to ensure safety and security. That must be the rule in dealing with events of very low probability but potentially catastrophic consequences.

Summary of Conference Papers

The following summary of papers prepared for the conference is organized in four segments, just as the papers were presented and as they are reproduced herein. The focus of these papers is on experience and issues with:

- Nuclear weapons
- Nuclear power plants
- The regulatory process
- The formation of public perceptions of all aspects of the nuclear enterprise

Session I Safety Issues—Nuclear Weapons

As long as nations maintain nuclear arsenals, it will be essential to continue efforts to reduce the risk of accidents. These efforts will require diplomatic means to reduce the number of weapons and prevent their proliferation as well as technical means to improve the security and operational safety of such weapons.

The safety record of the U.S. nuclear weapons program since 1945 is superb; there has not been a single casualty directly caused by an accident or incident leading to the explosion of a nuclear bomb. In the early years of the program, safety was not viewed with the same urgency as were modernization and design improvements to meet military requirements. Moreover, in those years much less was known about radioactive fallout. There were few analytical tools and limited capabilities to simulate the effects of fallout, particularly at large distances. Fortunately, the priority of safety struggled to the fore in the nuclear weapons enterprise during the 1970s and 1980s. Today, however, we are still working our way through safety challenges that were made more difficult by design decisions taken to meet military requirements set during the height of the Cold War.

Two papers presented by **Sidney Drell**[2] and **Robert Peurifoy**[3] addressed the technical improvements that have been incorporated into U.S. nuclear bombs to ensure their safety to the highest degree practical, and illustrated how they removed vulnerabilities that were learned from several serious "close call" accidents. They also documented the critical role played by a small cadre of courageous leaders in the weapons program. Aided and abetted by outside political and public support for those concerned about safety, these leaders were persistent in their efforts to overcome resistance within the nuclear enterprise to recognize the problems and to expedite the implementation of necessary design changes.

Efforts are ongoing to increase our confidence in being able to assess accurately the likelihood of accidents occurring as well as their consequences.

The formal methodology of probabilistic risk assessment (PRA), as described in **Christopher Stubbs's**[4] paper, is generally relied on for estimating the likelihood of situations occurring that could lead to accidents. It is a highly developed analytical approach that faces its most demanding challenge when addressing problems that have low probability but high consequence for failure. The methodology involves complex systems and a combination of sophisticated hardware and software as well as humans in short-time-scale critical decisions. PRA also needs solid data as important ingredients in its calculations to ensure that it is dealing with the real world, not just the world of mathematics, and that it can describe the situation for both the weapons and the power operations in the nuclear enterprise.

The normal, or Gaussian, probability distribution, which is frequently used because of its mathematical simplicity, has narrow tails that can be misleading in making confidence estimates. It is extremely difficult to model very-low-probability events since one is working far out on the tails of the probability curves where data is sparse. Thus it is important to provide strong support for technical research to gain a better understanding of the shape and extent of those tails based on data.

Discussion emphasized both the strengths and the limitations of PRA and its value in estimating risk assessments, both relative and absolute. The former (relative) is most useful for guiding design and investment choices in risk reduction. The latter (absolute) faces a more difficult, and less clear, challenge for guiding policy decisions. These considerations were applied in reviewing the Challenger space shuttle and Fukushima accidents to help understand the strengths and limitations of PRA.

With the current proliferation as well as the potential for continued proliferation of nuclear weapons, particularly in regions of political and strategic instability, a new global danger is developing: radioactive fallout at large distances caused by limited numbers of nuclear explosions in a regional conflict or terrorist incident.

As discussed by **Raymond Jeanloz**[5] in his paper, this problem is tough to analyze not only because of uncertainties in the source (the nature of the incident and number of bombs exploded, bomb designs, yields, and altitudes of burst), but also because of large uncertainties in the transport of radioactive particulates in the atmosphere over global distances. Notwithstanding these uncertainties, based on the considerable information gained from previous nuclear weapons test explosions and accidents at nuclear power plants that produced fallout at great distances, the damaging health effects could greatly exceed those from Chernobyl—which, as the largest reactor accident, put several megatons of debris into the atmosphere.

Whether arising from nuclear detonations in a regional conflict or from a reactor explosion, fallout at large distances will have a serious societal impact on the nuclear enterprise. There is little understanding of the reality and potential consequences of such an event if it were to occur halfway around the world. An effort should be made to prepare the public to face the impact of such an event, including providing information on how best to respond.

In his paper discussing the outstanding safety record of U.S. naval nuclear power that has operated for more than a half-century without a major safety incident due to nuclear failure, **Drew Dewalt**[6] reviewed key operating principles stated by Admiral Hyman Rickover, former director of the Naval Reactors Branch in the Bureau of Ships. They include strong central technical control, technical competence, conservatism of design, compliance with detailed operating procedures, and establishment of a safety culture. Each of these principles is also relevant to both the nuclear weapons and civilian power programs.

It is necessary, however, to acknowledge inherent differences in applying these key guidelines to the civilian nuclear power enterprise. Differences include the need for civilian utilities to control costs while remaining economically competitive, the existence of a wide diversity of reactor designs, and formal regulatory oversight. These issues were the focus of discussion at the next two segments of this conference.

Session II Nuclear Reactor Safety

Ed Blandford and **Michael May**[7] emphasized the importance of taking seriously lessons learned from previous nuclear reactor accidents. Regrettably, the Fukushima accident revealed significant shortcomings in the international commitment and ability to learn from previous experience and to implement appropriate safety measures. There were important precursors to safety failures in all major accidents, including Fukushima. For example, a major flooding incident in France should have alerted Japan to the vulnerability of locating the back-up power system at Fukushima below ground.

One important lesson that should be learned from Fukushima is that there were inadequate means of verifying the conditions at crucial areas of the plant following the power failures. This lack of knowledge delayed appropriate response strategies that could have prevented the partial meltdowns of the reactors. Another important lesson from Fukushima is that preparing for rare events that may occur only once in a thousand years is not an unreasonable criterion when dealing with an enterprise involving a large number of units for which a failure can have catastrophic consequences.

Previous nuclear reactor accidents also demonstrated the importance of setting up a command structure with clear lines of responsibility and communication with carefully prepared plans to manage events following an accident. The structure should encompass all relevant actors: government officials, independent regulators, management, and first responders.

Almost all of the currently operating reactors—approximately 400 worldwide, including approximately 100 in the United States—are second-generation models that lack modern safety features. Technical developments being incorporated in third-generation and more advanced nuclear reactors are impressive, as described in the paper prepared by **Per Peterson** and **Regis Matzie**,[8] and have resulted in improved efficiency and safety of operations. INPO, which was created in 1979 after the Three Mile Island accident, is providing a well-balanced process with a combination of transparency and privacy for monitoring safety improvements and more fully incorporating the lessons of previous accidents and close calls in the U.S. reactor enterprise. Unfortunately, WANO, which is the global version of INPO, is less effective because it lacks teeth for comparable enforcement authority, and carrots to offer financial incentives. This deficiency requires a particularly urgent remedy as the majority of reactors providing power around the world are approaching the end of their lifetimes. The focus should be on monitoring the aging of materials and the adequacy of the safety measures being implemented in these older reactors.

With strong government support, significant advances in safety can be achieved and are being introduced into modern designs—generations three and four—that incorporate more redundancy and passive methods for ensuring safety and more rugged modular designs. There remain, however, issues of cost, not only of up-front construction but also of insurance against liabilities in the event of an accident that releases radioactivity. How these improved-safety-versus-cost issues in the nuclear sector will balance out economically against progress being made in alternatives for providing clean energy remains to be seen, posing a challenge to the future of the nuclear energy enterprise.

Robert Budnitz[9] discussed what to do with the high-level radioactive waste generated primarily by civilian nuclear power reactors. This has been a contentious issue for decades that concerns both permanent (that is, for millennia) disposal of radioactive waste in deep underground caverns providing safety against theft or environmental damage; and interim storage (from decades to a century) in safe, secure havens on the earth's surface. In both cases, it is generally agreed that the issues under debate are not technical but political, and that costs are a relatively minor factor.

William Martin and **Burton Richter**[10] considered three scenarios to face the challenge of internationalizing the process of ensuring reactor safety around the world. Business as usual seems unnecessarily risky. It worked fairly well for the last forty years as long as nuclear power was confined to that group of developed countries that more or less worked together to establish the current patchwork of nuclear safety and regulatory provisions. But it may not do very well with the new players—primarily the developing countries—where the technical base and safety culture are not as strong.

A second scenario is to broaden somewhat the international conventions, including those generated by the IAEA. Internationalizing safety standards and their enforcement, particularly for the more hazardous aspects of operations, will prove to be an especially difficult political and economic challenge. The nuclear power industry is a commercial profit-making enterprise with economic pressures to control costs while simultaneously trying to establish common safety standards. One must also consider additional responsibilities: ensuring the security of reactors against sabotage or terrorist attack; guarding against the spread of technologies that could contribute to the proliferation of nuclear weapons capabilities; and providing for the safe transport of nuclear materials from on-site spent fuel water ponds to interim dry cask storage. Such arrangements must meet the possible concern of each newly participating nation that its prestige and potential emergence as a nuclear nation will not be diminished.

A third and more radical idea would be to establish an international nuclear bank to take responsibility for both the reactor and the land, which could be designated as international territory and given the same protection as an embassy. The reactor itself would be leased, and the electricity even subsidized in compensation to the participating nation for giving up part of its sovereignty over the plant itself.

In his paper, **Jeremy Carl**[11] addressed the current status of the nuclear power enterprise and concerns about its future, including the ambitiously planned expansion of nuclear power in the world's two most populous nations, India and China. India, with a fairly poor safety record during the 1990s, is planning to build the world's largest nuclear power station: ten gigawatts under contract with AREVA, with a planned total of sixty-three gigawatts by 2032. In the wake of Fukushima, not to mention the enduring shadow of the Bhopal Union Carbide tragedy, these plans are generating considerable political controversy within India. The plans also must contend with liability laws that do not conform to international standards, providing barriers to foreign investors.

China is a very ambitious player, with plans to commission forty gigawatts of incremental nuclear power by the year 2015. There have been allusions to its lack of trained regulatory personnel and engineers for such a large task, but little is known with confidence because of tight security. However, there has been solid evidence of accidents and costly safety failures. In particular, there was a relatively serious accident in 1998 at China's first indigenously designed reactor, located just sixty miles from Shanghai, which caused the plant to be closed for almost a year.

Evidently there is cause for concern about the frailty of the global nuclear enterprise.

Session III Economic and Regulatory Issues

Regulatory issues were the focus of the conference's third session. An issue of major importance is establishing and maintaining an independent authority protected from regulatory capture by either the government or the industry being regulated.

As a commentary on the paper submitted by **Gary Becker,**[12] who was unable to attend the conference, **John Cogan**[13] surveyed developments over time that have influenced the effectiveness of the regulatory process, noting that the practice of capturing regulatory agencies dates back to the creation of those agencies. One of the earliest examples is the Interstate Commerce Commission, established in 1887 by the Grover Cleveland administration, in which the attorney general just happened to be a former railway man.

The U.S. nuclear regulatory system, administered by the NRC and the Federal Energy Regulatory Commission (FERC) and operating with appropriate checks and balances, is praised as the gold standard of the world. A key reason for this is that the regulatory structure is much more balanced on performance than on compliance regulations. This is in sharp contrast to what was learned to be the case in Japan after the Fukushima incident.

John Taylor and **Frank Wolak**[14] contrasted financial regulation with nuclear power regulation. They emphasized the harmful effect of failure to enforce existing regulations against excessive risk-taking in the financial markets versus the challenging complexity of adhering to a broad range of worldwide norms and standards applied to a wide array of technologies in the nuclear power industry.

The effective administration of the NRC and its reliance on independent scientific inspections and study panels have been valuable in maintaining competence and avoiding regulatory capture by either the government or the operators, a problem that has troubled the world of financial regulation. To be sure, the system is not completely free of the danger of regulatory capture by industrial interests because there is considerable movement of personnel between the regulatory agencies and the firms they regulate. In addition, a high percentage—90 percent—of the regulatory fees supporting the NRC are paid by the industry being inspected. By squeezing that budget, industry can reduce the scope of inspections by the NRC and the technical work of its research office, thereby reducing its clout. A big step downward has already occurred over the past twenty years as the result of a cutback in this support by a factor of five. There is concern in the United States that constant vigilance is necessary to ensure the continued independence of its nuclear power enterprise. This attitude is also a valuable model for the establishment of nuclear power regulation globally, particularly in light of the failures of oversight that led up to the Fukushima accident.

The major financial barriers looming ahead for the nuclear power enterprise appear to be in the form of economic realities: up-front financing of construction, liability insurance, and litigation costs stemming from public concerns about the consequences of a nuclear accident. Additional challenges to the economic competitiveness of nuclear power are being posed by increasing support for innovation in alternative forms of clean renewable energy sources that offer the promise of reducing costs, and by the surge in the availability of natural gas and the reduction in its price.

In this climate, the future role of nuclear power is increasingly being debated worldwide, and particularly in the United States, which has not started construction of a new nuclear reactor since 1977.[15] In his paper on the economics of nuclear energy, **Michael Boskin**[16] reminds us that many unpredictable factors will determine whether nuclear power will grow from today's 6 percent share of total global energy to 8 percent, as has been predicted, or whether eagerly anticipated game-changers will rule the day.

In the meantime, however, certain basic factors will be instrumental in defining the near-term future of the nuclear enterprise. As the reactors age, they must be managed and cared for in much the same way as nuclear weapons are, with a life-extension program that installs important safety, security, and performance upgrades. Regulatory measures should focus on major safety issues while recognizing the importance of not interfering unduly with requirements for economically competitive operations.

Above and beyond formal procedures, the value of assembling an A-team of individuals committed to honoring the culture of safety that has proved so important to the nuclear weapons enterprise cannot be overemphasized.

Session IV Media and Public Policy

The fourth session explored how insights gained from the preceding sessions might be used to inform and raise public and governmental discussions regarding the nuclear enterprise in the energy and weapons arenas.

Educating the public to understand nuclear risks and consequences is an essential part of preparing a response to any nuclear mishap. In his introductory remarks, **Jim Hoagland**[17] pointed out that substantial effort has been made by the media to cover nuclear proliferation, the diplomacy around the issue, and stockpile safety. However, with a few exceptions, media attention does not focus on the enterprise that produces nuclear energy, its safety standards and records, and the regulatory process—until disaster strikes.

The implications of this deficiency were emphasized in **David Hoffman's**[18] paper on media coverage of the three major incidents involving nuclear reactors: Three Mile Island in 1979, Chernobyl in 1986, and Fukushima in 2011. Hoffman detailed how the polling data taken shortly after Fukushima revealed significantly reduced public support in many nations for building more nuclear power plants, and growing opposition to building them in one's own community.

Hoffman also highlighted the need for better communication among government and industry officials, both before and during a nuclear incident. Greater commitment by the media to participate in training exercises, disaster drills, and seminars organized by public authorities and reactor operators would help to develop expertise and confidence and would improve the quality of reporting. The media also should be more aggressive in establishing accountability and in examining and challenging industry, regulators, and the government about what should be—but is not being—anticipated. This is a big challenge in a free society, and it is even bigger in countries without a free or developed press.

In discussing the roles of the media, governments, and industrial leaders, two points deserve emphasis. In Hoffman's words, "There's a temptation to say after an event like Fukushima that the public is ignorant and irrational: *If only they had the facts, they would understand.* By this rationale, what's needed is simply to provide more factual information about nuclear power, and public opinion will improve. While such an assumption is logical, there are other factors. Public opinion is also profoundly influenced by subjective feelings such as fear, the unknown, and loss of control. These must also be taken into account when considering attitudes in the future, and the news media play a very large role in how the public shapes such fears."

"Training and improved understanding of science would make some journalists better," Hoffman concludes. "But this is not enough to transform public opinion. The credibility of the nuclear enterprise rests on the actions of governments, regulators, and utilities as well. They must absorb the lessons of Three Mile Island, Chernobyl, and Fukushima: a nuclear emergency is not only a crisis of information, but of confidence. Trust is fragile, and to break it once is to lose it for a long time."

At the core of this issue, one must always be cautious to heed the warning expressed by the physicist Richard Feynman in his summary of the Challenger shuttle accident in 1986: "For a successful technology, reality must take precedence over public relations, for nature cannot be fooled."

Notes

1. Shultz, George P., William J. Perry, Henry A. Kissinger and Sam Nunn, "A World Free of Nuclear Weapons," *Wall Street Journal* (January 4, 2007); Shultz, George P., William J. Perry, Henry A. Kissinger and Sam Nunn, "Toward a Nuclear-Free World," *Wall Street Journal* (January 15, 2008); Shultz, George P., William J. Perry, Henry A. Kissinger and Sam Nunn, "Deterrence in the Age of Nuclear Proliferation," *Wall Street Journal* (March 7, 2011).

2. "Designing and Building Nuclear Weapons to Meet High Safety Standards," by Sidney D. Drell, chapter 1, this volume.

3. "A Personal Account of Steps Toward Achieving Safer Nuclear Weapons in the U.S. Arsenal," by Robert L. Peurifoy, chapter 2.

4. "The Interplay Between Civilian and Military Nuclear Risk Assessment, and Sobering Lessons from Fukushima and the Space Shuttle," by Christopher Stubbs, chapter 3.

5. "Long-Range Effects of Nuclear Disasters," by Raymond Jeanloz, chapter 4.

6. "Naval Nuclear Power as a Model for Civilian Applications," by Drew DeWalt, chapter 5.

7. "Lessons Learned of 'Lessons Learned': Evolution in Nuclear Power Safety and Operations," by Edward Blandford and Michael May, chapter 6.

8. "Nuclear Technology Development: Evolution or Gamble?" by Regis Matzie and Per Peterson, chapter 7.

9. "The Spent Fuel Problem," by Robert J. Budnitz, chapter 8.

10. "International Issues Related to Nuclear Energy," by William F. Martin and Burton Richter, chapter 9.

11. "Fukushima and the Future of Nuclear Power in China and India," by Jeremy Carl, chapter 10.

12. "The Capture Theory of Regulation," by Gary S. Becker, chapter 11.

13. "The Federal Regulatory Process as a Constraint on Regulatory Capture," by John F. Cogan, chapter 12.

14. "A Comparison of Government Regulation of Risk in the Financial Services and Nuclear Power Industries," by John B. Taylor and Frank A. Wolak, chapter 13.

15. On February 9, 2012, the NRC approved construction of two new nuclear reactors at the Vogtle plant in Georgia.

16. "Discussion Notes on the Economics of Nuclear Energy," by Michael Boskin, chapter 14.

17. "Media and Public Policy," by Jim Hoagland, chapter 15.

18. "The Nuclear Credibility Gap: Three Crises," by David E. Hoffman, chapter 16.

Session I

Safety Issues—Nuclear Weapons

Designing and Building Nuclear Weapons to Meet High Safety Standards

1

SIDNEY D. DRELL

The safety record of the United States nuclear weapons enterprise is quite remarkable. We have built, deployed, exercised, and dismantled roughly 70,000 warheads and about 10,000 launchers. During the sixty-plus years since 1950 there have been a number of accidents, including thirty-two acknowledged "Broken Arrows" accidental events leading to losses, disappearances or crashes that involved U.S. nuclear weapons and delivery systems, and could have resulted in serious consequences (see Table 1A.1 in appendix on page 39). The fact that not one single nuclear warhead has directly caused a casualty is quite an achievement. That achievement required a lot of hard work and reflects on the determination of principled people to do the right thing, often against organizational resistance. The next paper[1] will discuss several of the "close-call" safety incidents and the effort led by its author to remedy the problems that caused them. His role is a lesson for all involved in such activities in which failures in safety can lead to devastating consequences.

In thinking about questions of safety in such very high-consequence operations, and what it takes to achieve an excellent safety record, I give highest priority to four criteria. They are, in no particular order:

1. Set the priorities in the proper order.
2. Bring to bear the best available analytical tools to analyze and understand the risks and consequences of failure.
3. Enforce rigorous discipline and accountability at each step in the process.
4. "Red Team" the activities—that is, perform critical reviews by independent technical experts, including exercising systems to the point of failure—with good communication channels up and down the line between management and engineers.

All four of these criteria are critically important.

1. Set the priorities in proper order.

The nuclear enterprise did not automatically start that way. Initially we did not fully appreciate the impact and the amount of radioactive fallout generated in nuclear explosions above ground in the atmosphere. Based on what was known at the time,[2] the bombs dropped on Hiroshima and Nagasaki in 1945 were detonated at altitudes between 1,500 and 2,000 feet. The military motivation for those detonation altitudes was to maximize the distance away from the aim point at which overpressures of approximately two to three pounds per square inch would be generated to cause significant structural damage. Such altitudes were high enough to prevent the fireballs from reaching the ground where they would have dug up extensive amounts of debris, mixed with radioactive fission fragments, which would have caused considerably more radiation sickness casualties.

However, during the years following the war, the United States conducted about 200 nuclear weapons tests above ground at the Nevada

Test Site (plus about 130 in the Pacific Ocean) that created fallout. This ended with the negotiation of the Limited Test Ban Treaty (LTBT) in 1963, which forbade any nuclear yield testing except underground. In fact, the U.S. Atomic Energy Commission (AEC) found itself with conflicting responsibilities: to create an arsenal of nuclear weapons for the United States against a growing Soviet threat that included nuclear weapons, and at the same time to ensure public safety from the effects of radioactive fallout. The same conundrum faced the AEC with regard to civilian nuclear power generation. It was charged with promoting civilian nuclear power and also with protecting the public.

When it became widely known that these tests were responsible for introducing significant amounts of harmful radioactive elements into the food chain (including iodine^{131} and cesium^{137} in particular), a strong public reaction resulted. This practice ended with the LTBT in 1963, but its legacy still persists today and challenges the credibility of the U.S. government and, indeed, of many governments in protecting citizens' safety from the effects of nuclear accidents. The nuclear concerns were greatly enhanced by the Castle Bravo test of a U.S. thermonuclear weapon in 1954 at the Bikini Atoll in the Marshall Islands. Its yield of 15 megatons was more than twice the anticipated value, and so was the fallout that was driven in an unintended direction by the change in the wind pattern, causing casualties among the civilian population more than 100 miles downwind.

To summarize, setting the priorities means insisting that the burden of proof rests on proving that the system is safe, rather than being satisfied with lack of evidence that it is unsafe. It was exceedingly difficult to implement such a priority for the stockpile in the chilling environment of the Cold War and within a process that evolved gradually through those years, starting in the 1950s. Modernization and improvement programs for the weapons gave priority to meeting military requirements, such as achieving maximum yield-to-weight ratios for warheads and maximum payloads and ranges for missiles. Safety was, in general, not viewed with quite the same urgency. Moreover, in the earlier years we knew much less and had few analytical tools and limited capabilities for simulation. Fortunately, the priority of safety struggled to the fore in the nuclear weapons enterprise during the 1970s and 1980s, spurred on by the determined commitment of a small cadre of courageous leaders in the weapons labs and enabled by the development of more powerful analytic tools providing critical data.

We are still today working our way through safety challenges that were made more difficult by design decisions before the end of the Cold War that gave higher priority to military requirements. Safety issues for military systems in other nuclear weapon states, which we do not control and about which we are not fully apprised, are of concern in considering the future of the nuclear enterprise. As we were recently reminded all too clearly by the reactor incidents at Fukushima, new initiatives in civilian nuclear power around the world are also cause for concern.

2. Bring to bear the best available analytical tools to analyze and understand the risks and consequences of failure.

This requires performing experiments and acquiring data which provide a basis for understanding how to design weapons that meet attainable goals that we set for limits on the probability of an unintended or accidental detonation and on the maximum acceptable explosive energy released in such an accident.

Modern nuclear weapons in the U.S. arsenal have an array of several thousand technically sophisticated components. Figure 1.1 illustrates what I am talking about. The picture shows an array of the components of one of the bombs designed to be carried on the B-52 and B-2 bombers in our strategic force as well as on a number of aircraft in NATO. It is necessary to understand the warhead electrical system, along with the nuclear package containing the high explosive and the nuclear material, well enough so that a probabilistic risk analysis can be made.

Setting Standards

It is widely alleged, and I am aware of no contradictory assertion, that the standard for the maximum acceptable nuclear yield in an unin-

Figure 1.1 The B61 nuclear bomb in various stages of disassembly of its almost 6000 parts. The nuclear component is contained in the blunt metal cylinder near the upper middle.

tended detonation was originally set in the 1960s when it was decided to deploy nuclear bombs aboard the U.S. Navy's aircraft carriers. A Navy captain, upon learning that nuclear weapons were to be loaded aboard the carriers in a compartment in proximity to the engineering operations center, asked himself: What if one of the weapons unintentionally or accidentally detonated? How large a nuclear yield would endanger the continued operation of the ship if one of the detonators were triggered by an accident? Detonation of the fifty or so pounds of high explosives could cause serious but acceptable damage to the bulkhead. But the new danger posed by a nuclear bomb is the release of radioactivity once a fission chain reaction is initiated. Given realistic conditions—i.e., the proximity of the weapons storage room to the engineering operations center and the limited radiation shielding of the floors and bulkhead between them—the flux of neutrons produced during the fission would present the greatest hazard.

The captain calculated that, if the fission chain continued long enough to produce an energy equivalent to four or more pounds of TNT, the flux of neutrons into the engineering operations center would approach or exceed the threshold for causing immediate incapacitation of the members of the crew in the room. As a result, the ship would be essentially dead in the water. Fission neutrons are emitted typically with one to two MeV (megavolt) energy, and cannot readily be absorbed by iron or steel walls less than a foot thick. (The large flux of gamma rays emitted during the fission chain is more readily absorbed.) This was accepted as a sensible and practical criterion to design to. As a result, nuclear weapons were designed to ensure that the fission chain will terminate quite prematurely. More precisely stated, the criterion is to not initiate more than about 10^{17} fissions following an accident that triggers a detonator at any one point and that ignites the high explosive in the bomb.

Having established the criterion for safety in terms of its consequences—no nuclear energy release exceeding the equivalent of four pounds of TNT—it is also necessary to set a goal for a limit on the acceptable risk of failure in meeting this standard. Once that standard is set, the challenge is to do the experiments and get the data upon which to base a probabilistic risk assessment.

The standards set by the DOE and DoD can be summarized as follows:[3]

1. The probability of a premature nuclear detonation due to warhead component malfunctions in the absence of any input signal, except for specified monitoring and control ones, shall not exceed:
   a. 1 in 10^9 per warhead lifetime, for normal storage and operational environments
   b. 1 in 10^6 per warhead exposure to abnormal environments prior to receipt of a pre-arm or launch signal, such as a lightning bolt, or due to an accident

It was also specified that this safety shall be inherent in the nuclear system design.

2. The probability of achieving a nuclear yield greater than four pounds TNT equivalent shall not exceed 1 in 10^6 in the event of an accident that could create two nuclear detonation risks:
   a. The high explosive is insulted and detonates at one point, or
   b. The firing set is activated because of faults in the electrical system.

These numbers were judged to be achievable in practice. The important question of confidence in achieving them is discussed in Stubbs's paper.[4]

Thirty-five very low-yield experiments were authorized by President Eisenhower and performed "down hole" at Los Alamos during the 1958–1961 moratorium with the Soviet Union on nuclear testing. They were designed to slowly creep up to producing no more than several ounces of TNT equivalent yield. What we learned from these measurements helped to identify the so-called one-point safety problems associated with some of the nuclear weapons systems of that time. The need for remedial action was demonstrated and, following the end of that moratorium, those issues began to be addressed systematically. Further confidence in being able to achieve these standards for one-point detonation has been gained from a continuing program of experiments and, in particular, advanced simulations with supercomputers doing high-fidelity calculations on the weapons in the current stockpile. This is an ongoing, very important part of the stockpile stewardship program, including the weapons Life Extension Program that the United States now relies on in the absence of underground test explosions.[5]

To meet the severe criteria for risk limitation set by the DoD and DOE, extensive experimentation and design work has been done at the three national weapons laboratories: Los Alamos, Lawrence Livermore, and Sandia. The Sandia National Laboratory focused primarily on designing and validating the performance of the thousands of parts that are outside the physics package in order to assure the safety of the weapons' electrical system (see Figure 1.1). The idea was to enclose the weapon's primary—i.e., the fission stage that releases the energy necessary to ignite the secondary, or fusion stage, of a modern thermonuclear weapon (H bomb)—together with the detonation firing apparatus in a sealed container that could withstand severe insults such as fire and crushing. The enclosure contains no source of electrical energy. Access to the container is allowed only through two specially designed, physically stout, and thermally tolerant pulse-pattern-operated switches, the so-called strong links.[6]

The firing apparatus is designed to become inoperable before the container or its two strong links (which must transmit the electrical energy to detonate the warhead) fail in the event of an accident or warhead component malfunction that leads to a modest thermal excursion from the normal. This is accomplished by co-locating a weak link component which can be shown to predictably fail prior to the barrier or strong links losing their integrity in the event of catastrophically severe environments which would eventually breach the exclusion region barrier to the warhead or the strong links.

The two strong links are in series and are of different designs, in order to minimize the risk of common-mode failures. One strong link is operated by human intent. It is designed to receive a pattern of several dozen short and long pulses in order to close and allow an arming signal to pass. If it receives a wrong signal in the event of an accident, or due to hostile action, it will lock up the system and no arming signal can be transmitted.

The second strong link must receive unique pre-programmed features of the designated trajectory of the weapon system during delivery to the target; for example, a pattern of pre-programmed accelerations, whether delivered by missile or aircraft. Otherwise it too will lock up. Extensive testing is performed to establish a one-in-a-thousand failure rate over long periods of time for each of these two strong links, and since they operate independently in series, it is believed that this meets the 1 in 10^6 criterion. This technology is called ENDS (enhanced nuclear detonation safety). Confidence in meeting this standard is continually evaluated. There have been bumps along the way and troubling resistance in the enterprise to fixing or removing the systems that failed to meet the officially adopted safety goals, as Robert Peurifoy recounts in his paper in chapter 2.

Beyond the two safety issues I have described in which an accident or a system failure could result in the release of nuclear energy—i.e., one-point safety and the electric detonation system—there are other significant safety issues that I will touch on very briefly. One has to do with the choice of the high explosive that initiates the fission chain reaction by squeezing the plutonium to criticality. The second one has to do with the choice of the missile propellant.[7]

One type of high explosive, HE, detonates much more readily in a violent accident than a second type, called simply IHE or insensitive high explosive, which is much more likely to simply burn, or deflagrate. In the former case, with HE, the plutonium would be widely dispersed and, as a result of the high temperature of the explosion, highly aerosolized as micron-size particles that become embedded in the lungs when inhaled and cause cancer. In contrast, with the IHE, the deflagration creates much lower temperatures, which disperse the plutonium over smaller areas and create larger particulates that enter into the digestive system and are rapidly removed by natural processes.

Another important difference between HE and IHE is that the latter has approximately only two-thirds of the energy per unit weight relative to the HE. As a result of this difference, ballistic missiles that are constrained by size and weight, such as the modern Trident missile in the nuclear submarine force, were designed with HE in order to permit them to load more, heavier, and higher-yield warheads. This decision enhanced their military parameters at the expense of safety against plutonium dispersal in the event of a handling accident. Such decisions during the Cold War reflected the clash between added safety and military requirements, and today the Trident force reflects that choice. In contrast, warheads deployed on modern intercontinental ballistic missiles (ICBMs) and bombs deployed on aircraft are not limited by such stringent geometric constraints.

Similar choices were made for the missile propellant, between a high energy propellant and a composite version that is very much more difficult to detonate but that delivers approximately 4 percent less impulse per unit volume of fuel, thereby decreasing the maximum missile range by approximately 8 percent for a fixed volume of propellant. Modern ICBMs are powered by the safer composite fuel while the Trident submarine-launched ballistic missiles (SLBMs), with their restricted volumes, are powered by the former more detonable high-energy one.

3. **Enforce rigorous discipline and accountability at each step in the process.**
4. **Red Team the activities—that is, perform critical reviews by independent technical experts, including exercising systems to the point of failure—with good communication channels up and down the line between management and engineers.**

These two criteria are important in conducting operations in any high-quality manufacturing organization or construction project. They are essential when dealing with an enterprise such as we are discussing that has to meet such exacting standards because of the high consequences of failures. This applies not only to the nuclear weapon program but also, of course, to civilian nuclear power. Management, with line responsibility from top to bottom, must enforce a culture of safety and accountability throughout the process. The use of Red Teams—by which I mean a rigorous independent peer review process, by both internal and external teams operating independently—has been demonstrably very valuable in ferreting out dangers that may have been overlooked or not properly evaluated. I say this on the basis of personal experience, including government-requested studies by the JASON group of academic scientists.

The accident that destroyed the Challenger shuttle in its 1986 flight is an example of what can happen so tragically when these standards are not enforced. As was learned from the post-accident investigation of the Challenger accident, the failure probability of the space shuttle's engine was estimated to be on an order of one in one hundred, according to a NASA consulting engineer. According to engineers working at the Marshall Space Flight Center, it was estimated to be about one in three hundred. But according to a claim by NASA management that was made at that time, it was only one in 100,000, and this was in spite of the erosion in the O rings that had been observed in previous flights, but wasn't anticipated, understood, or resolved.

In the weapons enterprise there were similar safety issues back in the early 1980s in several weapons systems. They were challenged from within but were resolved only after hard years of effort and outside support, plus public exposure in the media. This will be illustrated in the next paper by Bob Peurifoy, who fought with determination to get the enterprise to fix the systems that lacked ENDS that would enable them to meet the stated goals for low probability of failure of the arming, fusing, and firing system. My highest compliment to Bob is to recognize him as the Admiral Rickover of this program for insisting on the high performance and accountability standard that Rickover established in the nuclear Navy, where it remains today, unparalleled. Bob was applauded for his efforts upon his retirement in 1991 after thirty-nine years at the Sandia National Laboratories working in nuclear weapons RDT&E (research, development, test, and evaluation) and as a fierce advocate for nuclear weapons detonation safety, including nine years as vice president responsible for weapon testing, nuclear safety, reliability, quality assurance, and military liaison activities. Bob was personally recognized by then Secretary of Energy James D. Watkins for "exceptional service to the United States and its national security," for his "candor in raising critical safety issues," and for his "strong advocacy for improving the safety of deployed weapons."

The value of such informed fighters—in the enterprise, in the government, and in the media—is enduring. We are still addressing safety challenges made more difficult by decisions in the design of several systems during the Cold War.

Appendix: Technical Issues of a Nuclear Test Ban

Sidney Drell and Bob Peurifoy

Nuclear Weapon System Safety

Extensive improvements in weapon design since 1945 have resulted in a versatile and powerful arsenal of weapons and delivery systems that

---

Previously published as Sections 3 and 4 in Sidney Drell and Bob Peurifoy, "Technical Issues of a Nuclear Test Ban," *Annual Review of Nuclear and Particle Science*, no. 44 (1994): 285–327.

incorporate military characteristics responsive to U.S. needs as formulated during the Cold War. These military characteristics include high-yield and light-weight warheads that can be delivered accurately over long distances and that are mounted on delivery vehicles—aircraft and missiles based on land or at sea—so that the overall weapon system is deployed ready for strike. Inevitably, tensions arise when balancing safety vs readiness requirements, and compromises must be made. In particular, established technologies that increase the yield-to-weight ratio of a nuclear warhead may result in a reduced margin of safety for the total weapon system. We explore these technical issues in this section.

Discussion of the safety of a nuclear warhead focuses on the primary or boosted fission stage. In modern U.S. weapons the pit of radioactive, fissile material—plutonium (^{239}Pu) and/or uranium (^{235}U)—containing a H^2/H^3 gas mixture is surrounded by a shell of high explosive. Upon properly phased detonation, it generates the nuclear yield and creates the high temperature (10^6 °C) required to initiate the fusion process in a multistage thermonuclear weapon. An accident could initiate detonation of the high explosive, which potentially could lead to significant nuclear yield and/or plutonium scattering. For the first of these two scenarios there are two possibilities:

1. The accident stimulates the weapon's arming and firing system, which causes the weapon to detonate. This concern is valid, but its evaluation does not require nuclear tests.
2. The high explosive is detonated by some method other than the weapon's arming and firing system. For this risk, we must examine two conditions: detonation at one point and detonation at more than one point. If detonation occurs at more than one point, one cannot confidently calculate or adequately test for the probability distribution of possible yields. On the other hand, the probability that an accident would cause a detonation at more than one point was thought to be vanishingly small in the context of agreed-to risk by the Los Alamos and Livermore National Laboratories, the AEC, the DoD, and the military services. This assumption is based on the fact that, when detonated at one point, all the chemi-

> cal explosive will be consumed in a few tens of microseconds. It was difficult to imagine an accident that would cause multiple detonation points during this short time span.[8] Another factor to consider is the possibility of multipoint initiation of weapons due to propagation of explosions in bunkers. For a detonation at one point, the problem resolves to the question: What is the probability of assembling a critical mass, given that the one-point detonation occurs at the most sensitive point on or in the chemical explosive? One-point safety nuclear tests were essential to the development of the codes used to evaluate configurations and risk probabilities. At first, these codes were two-dimensional at best, and by today's standards they were superficial and inadequate. As a result of considerable advancements in computing power, code sophistication, data bases and, in particular, development of three-dimensional codes of appreciable capability, our ability to examine one-point safety risks has improved greatly. The codes could not have been improved without the ability to conduct dedicated nuclear safety tests.

From 1945–1951, the fission bombs of the U.S. were designed with nuclear capsules that could be manually inserted and removed. The capsule was to be inserted while the aircraft was en route to the target and removed before landing if the mission was aborted.

Beginning in 1952, warheads were designed with mechanically inserted capsules so that bombs could be carried external to the aircraft and to facilitate the making of nuclear-capable first-generation missiles. This technique relied on an electrically operated screwjack that would be activated to insert the capsule in flight en route to a target and (for bombs) that was reversible prior to landing if the mission was aborted. Normally the capsule would be positioned in the weapon but offset from the center of the high-explosive sphere. As long as the screwjack's motor was not activated inadvertently or by an accident, such a configuration was thought to be nuclear safe, although a chemical explosion would result in dispersal of plutonium. This configuration is similar in principle to the binary concept for modern chemical weapons.

As noted above, in the mid-1950s the design of nuclear weapons advanced to the present-day concept of sealed pits, with the fissile material permanently sealed within the high explosive. The superior military characteristics of this design include the more efficient use of nuclear material, and hence less weight and volume, for a given explosive yield. The sealed-pit configuration also improves operational readiness while reducing handling operations by military and maintenance personnel. However, it places increased safety demands on the electrical system for detonating the warhead in the case of severe abnormal environments or accidents. These demands were cause for serious concern because the early electrical systems for detonating the warhead could result in a nuclear yield if triggered inadvertently or by an accident. Some of the early sealed-pit warheads also could produce an unacceptably large nuclear yield if the high explosive was initiated at a single point. Nuclear testing demonstrated this characteristic and allowed appropriate corrective actions.

At about the same time, the U.S. entered a period of heightened international tensions, especially with the Soviet Union, that lasted for three decades. In 1957 the Strategic Air Command (SAC) began operating with forces armed on ground alert, and in 1958 it instituted around-the-clock airborne-alert operations. The U.S. also initiated large deployments of nuclear weapons to Europe.

One consequence of the rapidly expanding arsenal, then widely deployed and on airborne alert, was a large increase in the number of accidents involving nuclear warheads and components. The official list of U.S.-announced accidents involving nuclear weapons numbers 32, of which 5 occurred prior to 1956 and 26 occurred between 1956 and 1968, when the U.S. ended the SAC aerial alert following serious and well-publicized accidents over Palomares, Spain in 1966 and at Thule Air Force Base, Greenland in 1968. Table 1A.1 briefly summarizes these accidents, which raised concerns about the unpredictability of the potential accident environments and the vulnerabilities associated with early electrical safing features of U.S. weapons. Other incidents have involved delivery systems, but without weapon damage, e.g., when the engine of a B-52 caught fire upon being started during a practice drill at the Grand Forks, North Dakota SAC base in 1980.

Table 1A.1 Summary of Accidents Involving U.S. Nuclear Weapons

Accident Number	Date	Location	Weapon Configuration[b] Assembled Weapons	Unassembled Weapons	Type of Accident	Nuclear Weapon Response: HE Response — HE Burn	HE Detonate	Contamination[c]
1	02/13/50	Puget Sound, WA	—	X	Jettison, 8000'	—	X	—
2	04/11/50	Manzano Base, NM	—	X	Crash into mountain	X	—	—
3	07/13/50	Lebanon, OH	—	X	Crash in dive	—	X	
4	08/05/50	Fairfield-Suisan AFB, CA	—	X	Emergency landing, fire	—	X	—
5	11/10/50	Over water, outside U.S.	—	X	Jettison		X	—
6	03/10/56	At sea (Mediterranean)	—	X	Aircraft lost	—	—	—
7	07/27/56	SAC Base	—	X	B-47 crashed into bunker	—	—	—
8	05/22/57	Kirtland AFB, NM	—	X	Inadvertent jettison	—	X	X
9	07/28/57	At sea (Atlantic)	—	X	Jettisons, 4500' & 2500'	—		—
10	10/11/57	Homestead AFB, FL	—	X	Crash on takeoff, fire	X	X (low order)	—
11	01/31/58	SAC base overseas	X	—	Taxi exercise, fire	X	—	X
12	02/05/58	Savannah, GA	—	X	Mid-air collision, jettison	—		—
13	03/11/58	Florence, SC		X	Accidental jettison	—	X	—
14	11/06/58	Dyess AFB, TX	X	—	Crash on takeoff		X	X
15	11/26/58	Chenault AFB, LA	X	—	Fire on ground	X	—	—
16	01/08/59	U.S. base, Pacific	—	X	Ground alert, fuel tanks on fire	—	—	—
17	07/06/59	Barksdale, AFB, LA	X	—	Crash on takeoff, fire	X (1/3)[d]	—	X (1/3)
18	09/25/59	Off Whidbey Is., WA	—	X	Navy aircraft ditched	—	—	—
19	10/15/59	Hardinsburg, KY	X	—	Mid-air collision, impact	X (2/2)	—	—
20	06/07/60	McGuire AFB, NJ	X	—	Missile fire	X	—	X
21	01/24/61	Goldsboro, NC	X	—	Mid-air breakup	—	—	—
22	03/14/61	Yuba City, CA	X	—	Crash after abandonment	—	—	—
23	11/13/63	Medina Base, TX	—	X	Storage igloo at AEC plant	X	X	X
24	01/11/64	Cumberland, MD	X	—	Mid-air breakup, crash	—	—	—
25	12/05/64	Ellsworth AFB, SD	X	—	Missile reentry vehicle fell	—	—	—
26	12/08/64	Bunker Hill AFB, IN	X	—	Taxi crash, fire	X (3/5)	—	X (1/5)
27	10/11/65	Wright-Patterson AFB, OH	—	X	Transport aircraft fire on ground	—	—	X
28	12/05/65	At sea, Pacific	X	—	Aircraft rolled off elevator	—	—	—
29	01/17/66	Palomeres, Spain	X	—	Mid-air collision, crash	—	X (2/4)	X (2/4)
30	01/21/68	Thule, Greenland	X	—	Crash after abandonment	—	X (4/4)	X (2/4)
31	Spring '68	At sea, Atlantic	X	—	Lost weapons	—	—	—
32	09/19/80	Damascas, AK	X	—	Missile fuel explosion	—	—	—

[a] *Source:* DoD in coordination with DOE. 1981. *Narrative Summaries of Accidents Involving U.S. Nuclear Weapons 1950–1980.*

[b] The term "assembled weapon" refers to either the separable nuclear capsule that was installed but was not in the bomb's pit or a sealed-pit type of weapon with the nuclear material integral with the HE subsystem. "Unassembled weapons" means that the separable nuclear capsule was not installed in the weapon or that only weapon components were involved. (The USAF press release for accidents 1–13 used the term "assembled weapon" for the above plus where a capsule was on the aircraft).

[c] Contamination from all accidents except 29 and 30 was low in radioactivity and highly localized in areas affected.

[d] In the parentheses, the first number indicates the number of weapons that had the named response, and the second number gives the total involved in the accident.

Over the years, concerns about the safety of nuclear weapons have received continuing attention. At times they have competed with established requirements for military characteristics in the modernizing of the U.S. nuclear arsenal. Safety requirements for nuclear weapons apply both to the weapons themselves and to the entire weapon system. For the warheads these requirements mean that design choices for the nuclear components and for the electrical arming system must meet the desired safety criteria. For the weapon system, one must not only select the appropriate design choices but also implement operational, handling, transportation, and use constraints or controls to satisfy the safety standards.

Safety of the U.S. stockpile is the shared responsibility of the DOE and the DoD, by explicit direction of the President. The quantitative nuclear weapons safety criteria in effect today were established in 1968 by Carl Walske, then chairman of the Military Liaison Committee, and are summarized as follows:

1. One-point safety criteria.
   a. In the event of a detonation initiated at any one point in the high-explosive system, the probability of achieving a nuclear yield greater than four pounds TNT equivalent shall not exceed one in one million (1×10^6).
   b. One-point safety shall be inherent in the nuclear design, i.e., it shall be obtained without the use of a nuclear safing device.
2. Warhead/bomb premature probability criteria.

   "The probability of a premature nuclear detonation of a bomb (warhead) due to bomb (warhead) component malfunctions (in a mated or unmated condition), in the absence of any input signals except for specified signals (e.g., monitoring and control), shall not exceed:

   Prior to receipt of prearm signal (launch) for the normal[9] storage and operational environments described in the STS, 1 in 10^9 per bomb (warhead) lifetime.

   Prior to receipt of prearm signal (launch), for the abnormal[10] environments described in the STS, 1 in 10^6 per warhead exposure or accident."

In addition, qualitative safety standards have been specified for all U.S.-deployed weapon systems as well as for nuclear explosives in DOE custody. These criteria are to be implemented in the design of nuclear explosives and nuclear weapon systems to guard against nuclear detonations and (in the case of nuclear explosives in DOE custody) the dispersal of harmful radioactive material due to accidents, natural causes, or deliberate, unauthorized acts. Four safety standards for nuclear weapon systems are stated in DoD directive 3150.2 (February 8, 1984):

1. "There shall be positive measures to prevent nuclear weapons involved in accidents or incidents, or jettisoned weapons, from producing a nuclear yield.
2. "There shall be positive measures to prevent DELIBERATE prearming, arming, launching, firing, or releasing of nuclear weapons, except upon execution of emergency war orders or when directed by competent authority.
3. "There shall be positive measures to prevent INADVERTENT prearming, arming, launching, firing, or releasing of nuclear weapons in all normal and credible abnormal environments.
4. "There shall be positive measures to ensure adequate security of nuclear weapons, pursuant to DoD Directive 5210.4."

In this directive a positive measure is defined as "a design feature, safety device, or procedure that exists solely or principally to provide nuclear safety." The draft of a revised DoD directive 3150.2 (July 7, 1989) amends this definition to "a design safety and/or security feature, principally to enhance nuclear safety."

A similar DOE directive on nuclear explosives added a fifth requirement with regard to dispersal of plutonium into the environment as formulated in the DOE 1990 policy statement 5610.10 (October 10, 1990):

"All DOE nuclear explosive operations, including transportation, shall be evaluated against the following qualitative standards (in the context of this Order, the word, prevent, means to minimize the possibility, it does not mean absolute assurance against):

1. "There shall be positive measures to prevent nuclear explosives involved in accidents or incidents from producing a nuclear yield.
2. "There shall be positive measures to prevent deliberate prearming, arming, or firing of a nuclear explosive except when directed by competent authority.
3. "There shall be positive measures to prevent the inadvertent prearming, arming, launching, firing, or releasing of a nuclear explosive in all normal and credible abnormal environments.
4. "There shall be positive measures to ensure adequate security of nuclear explosives pursuant to the DOE safeguards and security requirements.
5. "There shall be positive measures to prevent accidental, inadvertent, or deliberate unauthorized dispersal of plutonium to the environment."

The DOE order defines positive measures as

> "design features, safety rules, procedures, or other control measures used individually or collectively to provide nuclear explosive safety. Positive measures are intended to assure a safe response in applicable operations and be controllable. Some examples of positive measures are strong-link switches; insensitive high explosives; administrative procedures and controls; general and specific nuclear explosive safety rules; design control of electrical and mechanical tooling; and physical, electrical, and mechanical restraints incorporated in facilities and transport equipment."

These official criteria and standards have stimulated and guided the efforts to advance the design of nuclear weapons for the past 25 years. These efforts and their experimental and analytical validation led to the concept of a modern, enhanced nuclear detonation safety (ENDS) system against premature detonation. They also stimulated the development of an insensitive high explosive (IHE), which possesses a unique insensitivity to extreme, abnormal environments, and of fire-resistant pits (FRPs) designed to further reduce the likelihood of plutonium dispersal in fire accidents. Over time, these and other design features and new technologies have been increasingly, but not completely, incorporated to further enhance weapon system safety to minimize the largest safety risks identified by fault-tree analyses and probabilistic risk assessments of the STS for each system. These technical advances have permitted great improvements in weapon safety since the 1970s.

Nuclear tests have been required in order to validate the IHE and FRP improvements. At the same time, technical advances have increased the speed and memory capacity of the latest supercomputers by factors of 100 and greater. As a result, during the past six years it has become possible to carry out more realistic calculations in three dimensions to trace the hydrodynamic and neutronic development of a nuclear detonation. These new results, though still relatively primitive, show how inadequate, and in some cases misleading, the earlier two-dimensional simulations were in predicting how an actual explosion might be initiated and lead to dispersal of harmful radioactivity or even to a nuclear yield.

Technical arguments for continued underground testing must be judged in the context of our confidence in and understanding of the critical components of weapon design, the steps that can be taken to improve safety, the requirements to validate any design changes, and finally, the determination of "how safe is safe enough." We discuss these issues in the following subsections.

Enhanced Nuclear Detonation Safety

Modern nuclear weapons are detonated by exploding the detonators placed in the high explosive that surrounds the nuclear material. A firing set that electrically initiates all the detonators must be armed before it can be fired. In a typical firing set, a capacitor discharge unit (CDU) initiates firing of the detonators. To arm the CDU, a high-voltage capacitor would have to be charged to several thousand volts, typically produced by a transverter within the firing set. The transverter converts a 28-volt DC arming signal to AC voltage and then, through transformer action and rectification, back to high DC voltage. The transverter output charges the high-voltage capacitor, and at the time of intended weapon detonation the energy stored in this capacitor is switched into the nuclear-system detonators for initiating the weapon. Figure 1A.1 depicts the general concept.

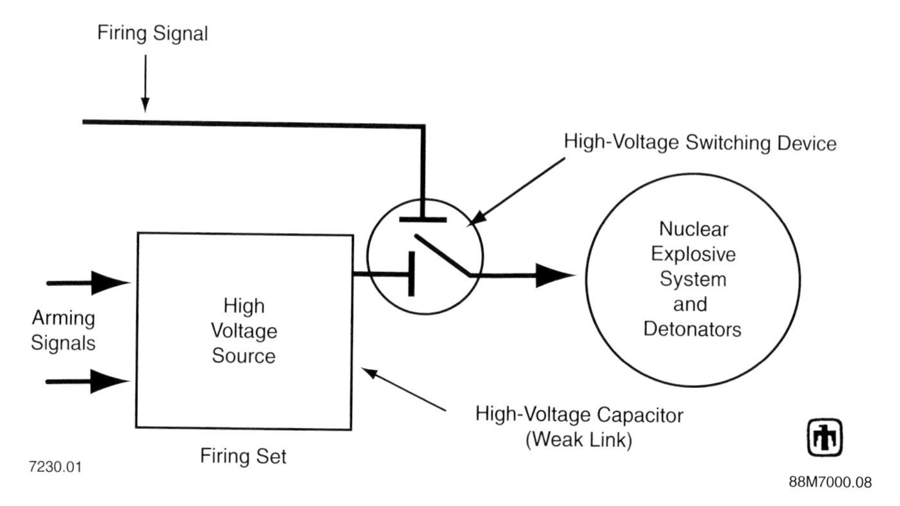

Figure 1A.1 Simplified basic configuration of the firing circuit used to detonate a nuclear weapon. Figure courtesy of Sandia National Laboratories Defense Programs and Surety Assessment Organizations.

The ENDS system is designed to prevent premature arming of nuclear weapons subjected to abnormal environments. Electrical elements critical to detonation of the warhead are isolated in an exclusion region physically defined by structural cases and barriers that exclude the region from all sources of unintended energy (see Figure 1A.2). The only access point into the exclusion region for normal arming-and-firing electrical power is through special devices called strong links that cover small openings in the exclusion barrier (see Figure 1A.3). Exclusion barriers may include diversion barriers designed to shunt unintended energy away from elements essential to detonate the weapon and/or insulation barriers designed to isolate elements essential for detonation from unintended energy. In addition to isolating the detonators from abnormal environments, essential elements for detonating the warhead are designed to become inoperable, i.e. to fail, before the isolation features fail.

The strong links are designed so that the probability that they will be activated by stimuli from an abnormal environment is acceptably small. In other words, strong links require an enabling input different from any electrical, mechanical, or environmental stimuli produced by exposure to an abnormal environment or accident (see Figure 1A.4).

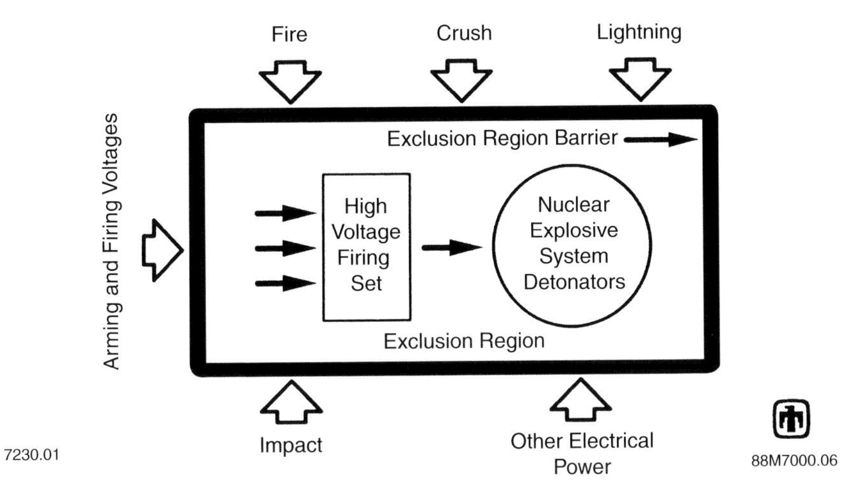

Figure 1A.2 Conceptual exclusion region used to protect critical firing apparatus. Figure courtesy of Sandia National Laboratories Defense Programs and Surety Assessment Organizations.

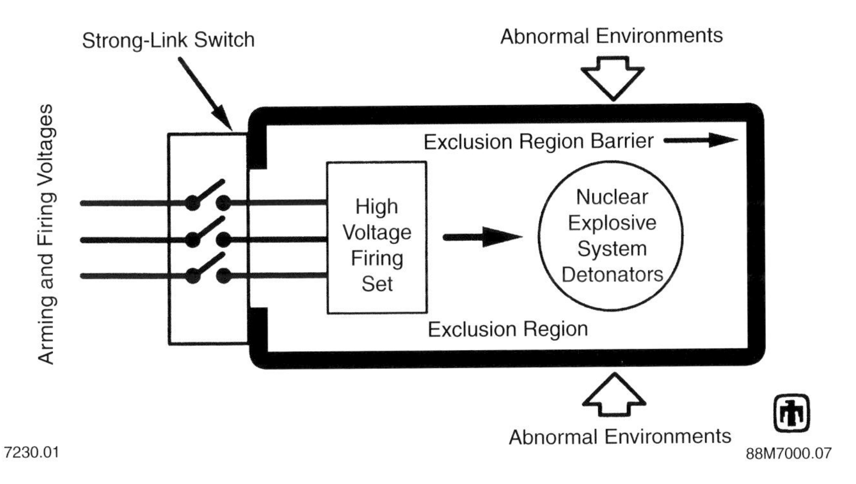

Figure 1A.3 Conceptual exclusion region entry protection. Figure courtesy of Sandia National Laboratories Defense Programs and Surety Assessment Organizations.

Over the years, detailed analyses and laboratory tests have strongly indicated that over a broad range of abnormal environments, a single strong link can provide isolation for the warhead at a probability of failure of less than 1 in 1000. However, prudence argues that to achieve

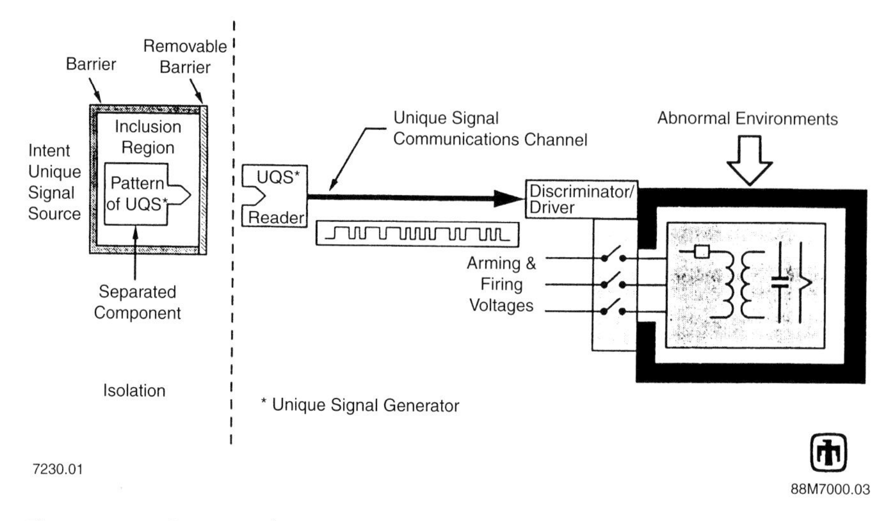

Figure 1A.4 Conceptual weapon system–level unique signal generation and communication channel. Figure courtesy of Sandia National Laboratories Defense Programs and Surety Assessment Organizations.

the safety requirement of a probability of less than 1 in 1,000,000 requires two independent strong links (of different designs to minimize the chance of common-mode failures) in the firing set. ENDS is designed to meet this criterion, i.e. both strong links must be closed electrically for the weapon to arm. Typically, one is closed by specific operator input, e.g. by the insertion into the aircraft electronics of a read-only memory chip containing the prescribed pulse-train information, and the other is closed by environmental input that corresponds to an appropriate flight trajectory (see Figure 1A.5). ENDS includes one or more weak links in addition to the two independent strong links, all enclosed in the exclusion region, to maintain assured electrical system safety at extreme levels of certain accident environments such as very high temperatures and crush. Safety weak links are functional elements such as capacitors (and the weapon high explosive) that are also critical to the normal detonation process. These weak links are designed to fail or become irreversibly inoperable in less stressing environments, e.g. lower temperatures than those that might bypass the strong links or cause them to fail. Calculations and

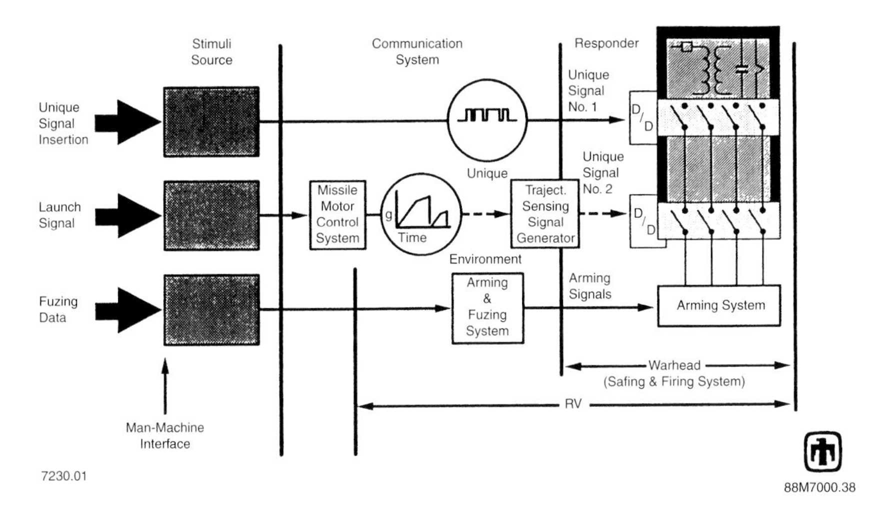

Figure 1A.5 Conceptual two-strong-link control and communication system. Figure courtesy of Sandia National Laboratories Defense Programs and Surety Assessment Organizations.

tests study the time-race between failure of the strong and weak links (see Figure 1A.6).

The ENDS system provides a technical solution to the problem of preventing premature arming of nuclear weapons subjected to abnormal environments. In concept, this system is relatively simple and lends itself well to safety assessments, tests, and production controls to assure that the ENDS design deployed meets the previously stated official weapon detonation safety criteria. The Sandia National Laboratories developed the ENDS system in the early 1970s and began introducing it into the stockpile in 1976. At present, more than three fourths of U.S. weapons are equipped with ENDS. According to current plans, all weapons without ENDS will be removed from the stockpile by the end of this decade. No nuclear testing is required to complete this important stockpile improvement program for enhanced detonation safety. However, new designs that might be deployed in the interest of higher confidence in command and control and, in particular, of better protection in the event of a terrorist seizure of a weapon would require limited nuclear testing to verify their effectiveness and reliability.

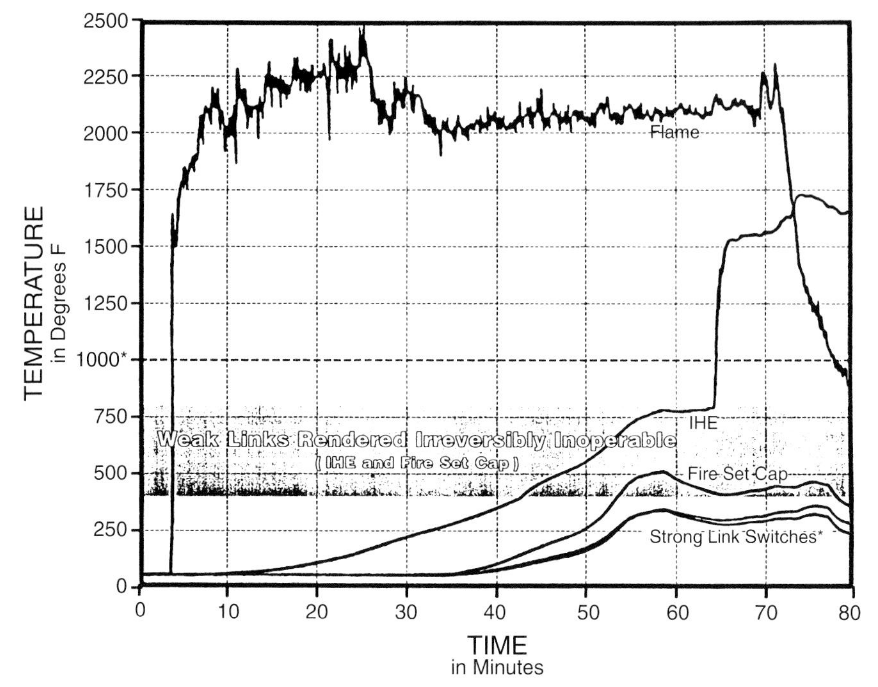

Figure 1A.6 Representative thermal environment test results for a weapon. Figure courtesy of Sandia National Laboratories Defense Programs and Surety Assessment Organizations.

Insensitive High Explosive

An insensitive high explosive (IHE) was developed to reduce the danger that an accident or incident would cause the high explosive surrounding the primary pit to detonate, causing radioactive contamination of the surrounding area and possibly a small nuclear yield as well. From the outset of the U.S. nuclear weapon program, designers have worked to reduce the probability of accidental nuclear explosions. Concerns about accidental explosions led the government in 1968 to adopt a formal set of safety criteria (listed on page 40). In addition to the probable catastrophic consequences of significant nuclear yield as a result of an accident, the lesser but significant consequence of plutonium dispersal was recognized. Toxicity of plutonium is far greater than that of any substance in previous experience, particularly if the plutonium is

Table 1A.2 The Hazards of Plutonium Exposure (Comparing Pu with Common Nuclides)

Nuclide	Sv per Bq[a]	REM/μCi
Pu-239	3.9×10^{-5}	144
Cs-137	3.5×10^{-9}	0.01
Sr-90	3.4×10^{-7}	1.26
^{3}H (water)	1.7×10^{-11}	6.3×10^{-5}

[a]Data from ICRP 1979, Report #30. Weighted by the relative cancer risks for each organ.

raised to a high temperature as a result of a detonation (in contrast to a deflagration) and is aerosolized into micron-size particulates that can be inhaled and become lodged in the lung cavity. Table 1A.2 illustrates the high degree of biological danger from the long-term α decay of ^{239}Pu compared with other less carcinogenic substances.[11]

The plutonium dispersal hazard, inherent in the sealed-pit designs of most U.S. nuclear weapons, gave priority to developing a high explosive that also had a very high energy density (and thus was required in small quantity) as well as a considerable insensitivity to detonation as a consequence of a violent accident such as an airplane fire or crash. In the 1950s, with the advent of the sealed-pit design, an energetic and stable high explosive known as PBX-9404 was developed. This explosive was composed of 94% hexamine nitromene (HMX) ($C_4H_8N_8O_8$), 3% nitrocellulose, and 3% of a phosphate ester plasticizer. Both the fuel and the oxidant are in the same molecule in HMX. Once initiated, chemical reactions proceed at extremely fast rates on a time scale of submicroseconds. Hence they can sustain detonation waves—shocks propelled by the quick release of chemical energy at supersonic velocities, typically 8mm/microsecond—but only if the reactions are complete before dissipated rarefaction waves reduce below a pressure/time or energy threshold. These conditions, which are necessary to sustain a minimum balance of energy loss from the region in which the reaction occurs, depend on the diameter of the material. The minimum diameter necessary to support the reaction is known as the failure diameter.[12] Although PBX-9404 could not be processed safely under normal munitions factory procedures, this explosive proved reasonably safe to handle under closely controlled procedures and could be fabricated to close dimensional tolerances and with low lot-to-lot tolerances.

The immensity of plutonium-dispersal problems—in terms of both the hazards and the costs of cleanup—created by accidents such as those at Palomares and Thule added urgency to the development of PBXs based on TATB (2,4,6-tri-nitro-1,3,5-benzenetriamine). Although this explosive, known as IHE, was first synthesized around the turn of the century, it remained a laboratory curiosity until it was realized that it possessed a unique combination of relatively high explosive energy density and an extraordinary chemical stability that makes it uniquely insensitive to extreme, abnormal environments. Table 1A.3 illustrates the added stability properties and the safety advantage of IHE compared with ordinary high explosives such as PBX-9404. IHE can be impacted into rigid targets at velocities exceeding 1500 feet/sec without provoking the release of substantial chemical energy. Traditional explosives release most of their chemical energy on impact at velocities on the order of 100 feet/sec. It is generally believed that the detonations that led to plutonium dispersal in the Palomares accident would not have occurred if those early warheads had been equipped with IHE. Although IHE designs are a significant improvement, they do not alleviate all safety concerns. For example, accidents may scatter some level of fissionable material simply as the result of mechanical breakup of the weapon.

In contrast to these highly desirable safety advantages, IHE is needed in greater weight and volume to initiate the detonation of a nuclear warhead because it contains, pound for pound, only approximately two-

Table 1A.3 Comparison of High Explosives

Relevant Properties	Conventional HE	IHE
Minimum explosive charge to initiate detonation (oz)	~10^{-3}	>4
Diameter below which the detonation will not propagate (in)	~10^{-1}	~1/2
Shock pressure threshold to detonate (kilobars)	~20	~90
Impact velocities required to detonate (ft/s)	~100	1800–2000

thirds of the energy of conventional high explosive. Therefore, the choice of IHE or high explosive in designing a warhead has an impact on the military characteristics of the weapon system that deploys it. The 1983 decision to design the W88 warhead for the U.S. Navy's new Trident II weapon system with conventional high explosive was based on operational requirements during the Cold War that were judged to override what were perceived as relatively minor safety advantages of IHE. This Trident safety issue was reviewed and new questions posed in the 1990 report of the Nuclear Weapons Safety Panel. For more detail, see Drell, Foster, and Townes, "Nuclear Weapons Safety Panel House Armed Services Committee,"[13] from which the following paragraph is excerpted to illustrate the effect of the choice of IHE vs high explosive on the military characteristics of Trident II:

"A major requirement, as perceived in 1983, that led to the decision to use high explosive in the W88 was the strategic military importance attached to maintaining the maximum range for the D5 when it is fully loaded with eight W88 warheads. If the decision had been to deploy a warhead using IHE, the military capability of the D5 would have had to be reduced by one of the following choices:

1. retain the maximum missile range and full complement of 8 warheads, but reduce the yields of individual warheads by a modest amount.
2. retain the number and yield of warheads but reduce the maximum range by perhaps 10 percent; such a range reduction would translate into a correspondingly greater loss of target coverage or reduction of the submarine operating area.
3. retain the missile range and warhead yield but reduce the number of warheads by one, from 8 to 7."

Replacing warheads with high explosive with new systems with IHE is at present perhaps the most effective way to improve safety of the weapon stockpile by reducing the danger of scattering plutonium. An understanding between the DOE and the DoD signed in 1983 calls for the use of IHE in new weapon systems unless system design and operational requirements mandate use of the higher energy (and thus the smaller mass and volume) of conventional high explosive. Moreover, the Senate Armed Services Committee in 1978 under Chairman John Stennis strongly recommended that "IHE be applied to all future nuclear weapons, be they for strategic or theatre forces." Although IHE was first introduced into the stockpile in 1979, by 1993 little more than one-third of the stockpile was equipped with it because most U.S. warheads currently deployed are those in the Trident I and II missile systems. The same is true for the stockpile envisaged under Strategic Arms Reduction Treaty (START) reductions. These warheads are denoted W76 for the low-yield Trident I version and W88 for the high-yield version developed for Trident II.

Because IHE and high explosive have different detonation characteristics, a redesign of the Trident warheads to replace high explosive with IHE would require an underground test program, albeit a modest and limited one. Details of the test program would depend on how the warheads were redesigned, i.e. whether a warhead with IHE currently deployed in other weapons was adapted to the Trident system or whether the pit of a retired warhead was rebuilt with IHE and adapted to the Trident.

Differences between high explosive and IHE are sufficiently large that one cannot rely on simulations and computer modeling alone to establish the effectiveness of a warhead thus altered. Much can be learned from the so-called hydro(dynamical) tests that analyze the chemical detonation and its impact on an inert nonfissioning pit. However, such tests, which would be allowed under a comprehensive test ban (CTB), would not estimate with sufficient confidence and accuracy the reflected shock waves, the final compression of the metal, and the condition of the boost gas. These are important in the analyses of the implosion and in determining whether the yield of the primary explosion is adequate to trigger the secondary explosion. Therefore, several underground tests would be required to confirm a new configuration with IHE and to establish that the primary has the designed characteristics and energy content.

Fire-Resistant Pits

When examining ways to enhance weapon safety, replacement of the high explosive with IHE should be considered together with the configuration of the rest of the weapon system. Risk analyses of the accidents that might occur during normal handling procedures and that could trigger an explosion are also needed. An accident might take the form of a dropped missile, which would lead to direct detonation of the warhead high explosive or the missile propellant, which in turn would initiate explosion in the warhead. A fuel fire on a bomber, provoked by an accident, an engine malfunction, or human error, might threaten the integrity and safety of any weapons on board. In the fire-accident scenarios, nuclear weapons could be involved in a hydrocarbon fuel fire of such intensity and duration as to breach the pit and thereby disperse the plutonium as a result of combustion and entrainment of the plutonium oxide particles into the fire plume. This concern led to the development of fire-resistant pits (FRPs).

An FRP has a metal shell with a high melting point that can withstand prolonged exposure to a jet fuel fire without melting or being breached by the corrosive effect of molten plutonium. One possible example of a suitable substance is vanadium, which can withstand a temperature of 1000°C for several hours and prevent the dispersion of plutonium into the environment. However, current technology for FRPs would not provide containment against the much higher temperatures (~2000°C) created by burning missile propellant. FRPs would also fail in the event of detonation of the high explosive surrounding the pit and therefore contribute to safety principally when used in weapons equipped with IHE. Once again, the sophistication of nuclear weapons and the effect of an FRP on the details of the implosion process would necessitate several tests to verify the theoretical design if this technology were introduced into weapons that currently do not contain an FRP. At present, only ~10 percent of U.S. weapons are equipped with an FRP.

Before adopting FRP technology, one would also need to accurately assess how much its improved containment of plutonium against fire would contribute to the overall safety of the weapons. For warheads on aircraft with bombs and air-launched cruise missiles, FRPs would add protection against the type of event that could have occurred in 1980 at the Grand Forks, North Dakota SAC base when an engine that was started on an alert B-52 caught fire. Fortunately, during the several hours that the aircraft fuel burned, the wind blew the flames away from the fuselage. The weapons on board had no modern safety features. At present, with the end of the Cold War, the SAC (now StratCom) no longer maintains bombers loaded with nuclear weapons on alert. All of the weapons have been returned to their bunkers, which greatly reduces the probability of an accident that could result in plutonium dispersal. However, such an accident could still occur if elements of the strategic bomber force are alerted during times of tension. One must take this possibility into account when evaluating the importance of underground tests in redesigning such weapons with FRPs.

One can also envisage going beyond FRPs to minimize the danger of plutonium dispersal or of a nuclear yield by separating the plutonium capsule from the high explosive prior to arming the warhead. This concept is a familiar one in the world of binary chemical weapons. However, the U.S. abandoned this nuclear weapon design in 1968 when it posited safety criteria that required inherently safe design. Recent advances in weapon technology and, particularly, in the means of reducing the weight and enhancing the reliability of mechanical devices may make the binary concept more attractive when evaluated against military requirements in the post-Cold War world than it was in the past. However, a move toward such safety-optimized designs would be costly and would require a more extensive development and underground bomb-testing program than may be appropriate in a post-Cold War world. In any case, before any such development is initiated, its contribution to safety should be analyzed in detail.

The question of FRPs also arises in connection with the missile force, in particular the Trident missile, which loads the warheads in a through-deck configuration (illustrated in Figure 1A.7) in close proximity to the propellant and third-stage motor of the missile. In evaluating weapon safety and the need for testing, we next consider the missile propellant and its impact.

Missile Propellant

Two classes of propellants are in general use in the long-range ballistic missiles of the U.S. One is a heterogeneous mixture of

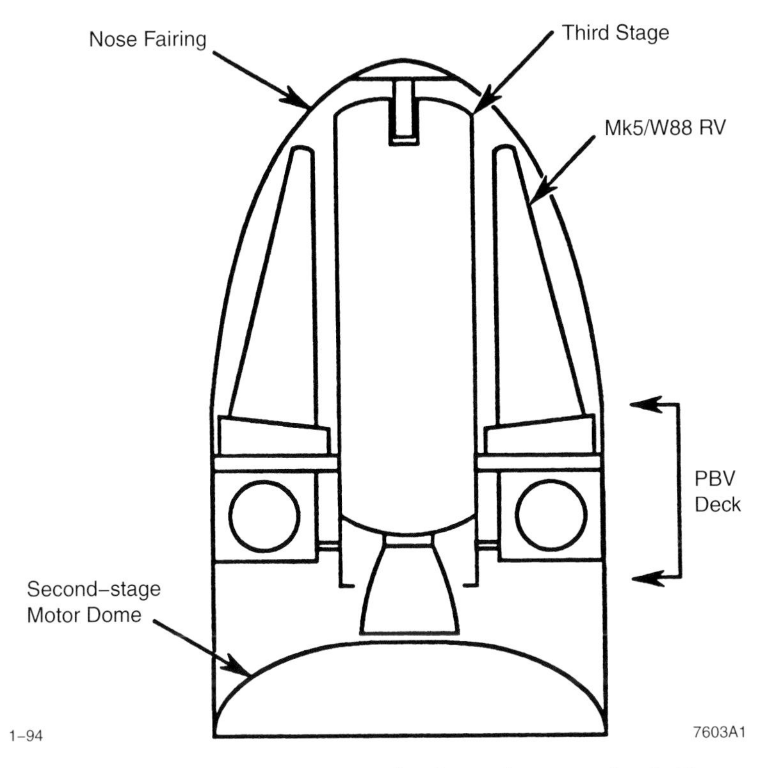

Figure 1A.7 Through-deck configuration for the Trident II warhead. Figure courtesy of Harvey & Michalowski.[14]

~70% ammonium perchlorate, 16% aluminum, and 14% binder. This composite propellant is dubbed 1.3 class. The other is a high-energy propellant dubbed 1.1 class. It is a cross-linked, double-base plastic made of ~52% HMX, 18% nitroglycerine, 18% aluminum, 4% ammonium perchlorate, and 8% binder. The 1.1 class denotes materials capable of mass detonation that can propagate from an initiation site to adjacent material or cause sympathetic detonation. The 1.3 class material is a fire hazard, but in contrast to the 1.1 class, it is much more difficult, if not impossible, to detonate. Table 1A.4 illustrates the important safety differences between these two classes of propellants. Of course, in addition to safety and stability of the material against shock-initiated detonation, the maximum specific impulse in a propellant is desirable in order to achieve long flight range. The maximum velocity achieved by a rocket at fuel exhaustion is proportional to the specific impulse for a given

Table 1A.4 Comparison of Missile Propellants

Relevant Properties	1.3 Composite	1.1 High Energy
Minimum explosive charge to initiate detonation (oz)	>350	~10^{-3}
Diameter below which the detonation will not propagate (in)	>40	~10^{-1}
Shock pressure threshold (kilobars)	No threshold established	~30
Specific impulse (s)	~260	~270

burn time, and the maximum flight range after rocket motor fuel exhaustion is proportional to the square of the velocity at burnout. Therefore, a 4 percent difference in specific impulse, as shown in Table 1A.4, translates to an ~8 percent range increase for the 1.1 high-energy propellant relative to the 1.3 composite propellant for constant burning times.

The choice of propellant raises the question of whether an accident during handling of an operational missile in transporting and loading might detonate the propellant, which in turn could cause a chemical explosion in the warhead. This explosion could then result in the dispersal of plutonium or worse—i.e. the initiation of a nuclear yield. This issue is a particular concern for the Navy's fleet ballistic missiles, the Trident I and Trident II, which are designed with through-deck configuration (illustrated in Figure 1A.7) to fit the geometric constraints of the submarine hull and at the same time achieve maximum range. In this configuration, if the third-stage motor were to detonate in a submarine loading accident, for example, a patch of motor fragments would impact on the side of the reentry bodies encasing each warhead. As a result, some combination of such off-axis multipoint impacts might detonate the high explosive surrounding the nuclear pit in one or more of these warheads and lead to plutonium dispersal or possibly a nuclear yield.

The 1990 Nuclear Safety Study Report[15] raised the through-deck configuration, together with the facts that the Trident warheads are designed with conventional rather than IHE and the rocket motor uses the detonable 1.1 class propellant, as a potential safety problem of the Trident force. The report questions whether the Trident warheads should be redesigned with IHEs and FRPs, perhaps with a buffer to shield them from the shock wave in the event of a third-stage detonation. Other possibilities were also considered, e.g. whether the propellant in the third stage should be changed to the nondetonable 1.3 class. This change would result in a range loss of no more than 4 percent, or a few hundred miles, because approximately half of the last velocity increment comes from the third-stage motor. Furthermore, when the START II reductions are implemented, the Trident force will carry fewer warheads—probably no more than four on a Trident II missile, which now carries a full load of eight. In this case a Trident II would have a maximum range no less than its present fully loaded configuration, even without a third-stage motor.

The Trident safety issues lead us to question how much money the U.S. wants to invest in further enhancing weapon safety and whether we want to continue underground testing in the years ahead. Changing the propellant in the third stage of the missile or removing the third stage altogether would require no yield testing of the warhead but would necessitate a development program of the weapon system at a cost that the Navy has estimated to be 1.6–1.8 billion dollars.[16] These cost figures apply if the high explosive in the Trident warheads is not changed. If the high explosive is replaced with IHE, the costs climb to 3.6–3.8 billion dollars for a START II–size force. To realize the full safety advantage of the change to a nondetonable propellant would require changing the high explosive in the warhead to IHE and adding an FRP.

Reliability of the Nuclear Weapon Stockpile

Figure 1A.8 illustrates the process of predicting and assessing the reliability of a U.S. nuclear weapon type. The initial step is the development of a reliability model of the weapon design, less the device. Figure 1A.9 shows an example of a top-level model for a bomb. This block diagram is greatly simplified and is intended only to illustrate methodology. To complete the model each major node is broken into more detailed models. For example, J_1 would be expanded to contain the safety switch and wiring. The safety switch in turn would be further expanded. In some cases the system designer chooses to include parallel (dual-channel) components to assure adequate reliability. For example, in Figure 1A.9,

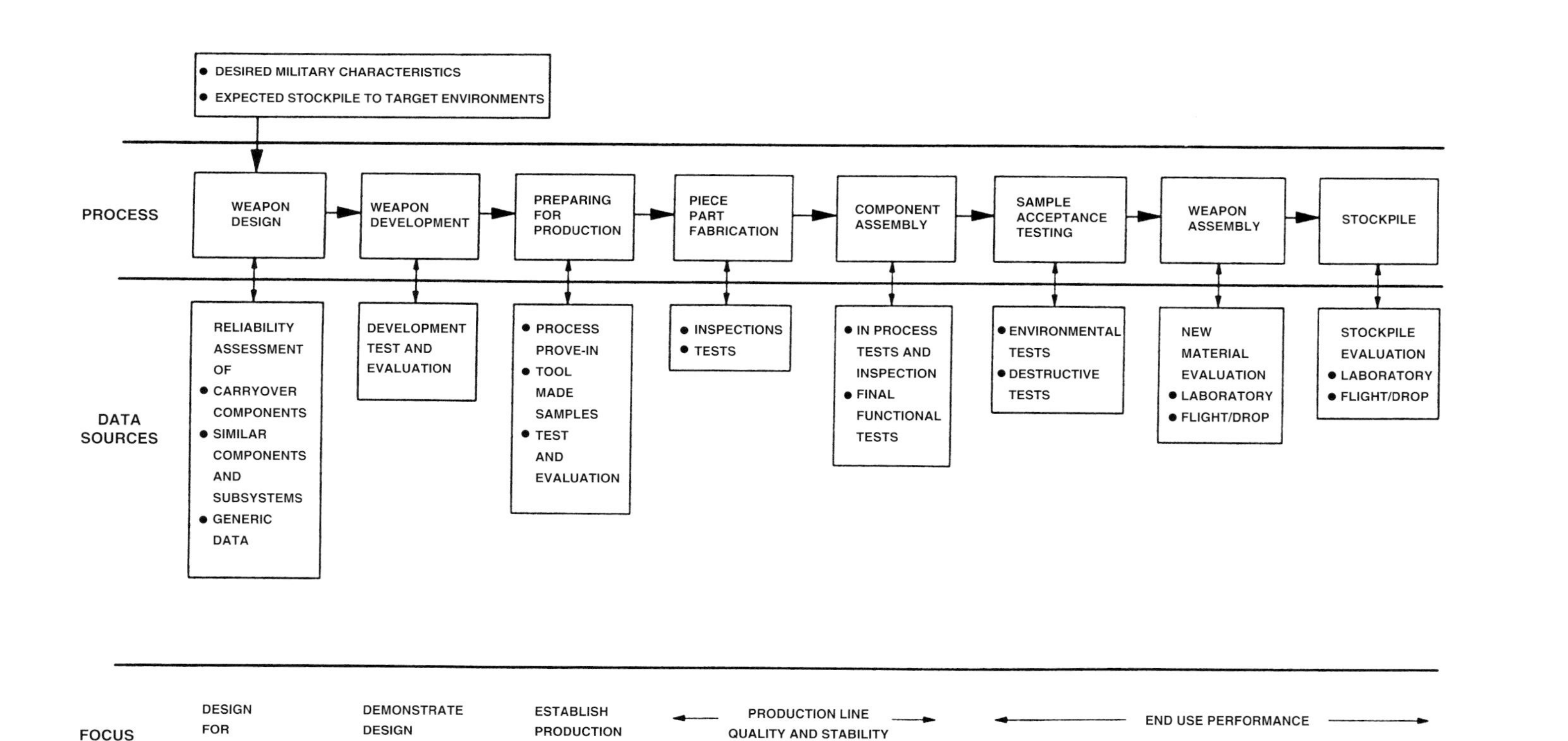

Figure 1A.8 Block diagram displaying reliability allocation/prediction/assessment process. Figure courtesy of Sandia National Labo-

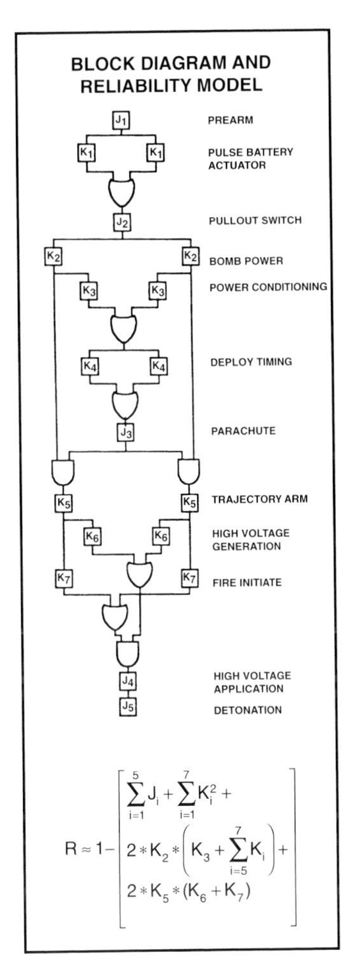

Figure 1A.9 Simplified bomb reliability block diagram. Figure courtesy of Sandia National Laboratories Defense Programs and Surety Assessment Organizations.

the thermally activated pulse batteries and main thermal batteries (bomb power, node K_2), are duplicated. Either (OR gate) will power the weapon if it was prearmed at release and if the pullout switch and pulse battery actuators function correctly. Some events require two properly timed inputs (AND gate). For example, the trajectory arm event (K_5) requires electrical power during a proper parachute opening—a voltage-deceleration-time profile.

This reliability model, which allocates a degree of reliability (actually, of unreliability) to each component, is used to estimate the overall bomb reliability. Furthermore, it provides the necessary detailed insight to determine where the dual-channel approach is effective in meeting the overall reliability goal or where redesign or increased emphasis on the reliability of a specific component has the greatest potential for overall increased reliability.

Components used in previous designs have an established production and stockpile history. If the stockpile-to-target environments are expected to be similar in a new application, the reliability allocation is straightforward, i.e. the existing data base is used. For new component designs, either the allocation is based on "looks-like" performance of similar components or, if there are no known "looks-like" components one relies on engineering judgment, and the component is flagged for special design and evaluation attention.

Once the component designers have agreed to the reliability allocations, test and evaluation programs are established to reasonably assure that the allocations will be met or, if not, that the shortfalls will be discovered. During development, this process results in refined or modified allocations and reliability predictions.

As the development program winds down, emphasis shifts to preparation for production and initial fabrication of piece parts (i.e. nuts, bolts, gears, levers, batteries, integrated circuits, resistors, capacitators, etc) and component assembly. The thrust of this effort is to establish a production capability of sufficient quality and stability to assure that the product reliability requirement will be met with homogeneous hardware of this type. During this development and preparation-for-production period, we refer to allocation and prediction, not assessment. The Sandia National Laboratories' weapon reliability and stockpile evaluation organizations hold that true reliability cannot be assessed without test and evaluation data from components and subsystems that could enter, will enter, or already have entered the stockpile. This information comes from three sources: component sample test programs, which test random samples of components designated acceptable for use in next assemblies; the New Material Evaluation Program, which examines, tests, and evaluates newly assembled weapons by tearing them down; and the Stockpile Evaluation Program, which draws random samples of weapons of each type (typically 11 per cycle—a cycle is every one or two years, depending on the accumulation of test results). Laboratory examinations and tests and realistically configured, instrumented drop/flight tests are conducted, the latter as part of military service evaluations of weapon systems. This multiphase program yields two useful results: (*a*) this-point-in-time reliability assessment, which draws on the cumulative reliability data bases, and (*b*) performance trends. The first is useful for real-time reliability assessments and feeds targeting algorithms. The real-time assessment is generally considered valid for future projections if no significant degradation is detected. The second is important for contemplating corrective actions in case a significant performance degradation is detected.

A Nondevice Component

An example of a nondevice component is the MC2969 strong-link switch, a safety switch used in early ENDS designs. The MC2969 was first used in bombs. The function of this switch is described in Section 3.1. The MC2969 contains 14 electrical contacts that, when closed, allow passage of all arming and firing signals into the exclusion region.[17] The switch is moderately complex and contains ~500 piece parts. Closure of the switch is accomplished by delivering a precise pattern of 29 long and short electrical pulses plus 10 operating pulses, typically by a bomb-prearming controller in the delivery aircraft. The pulse logic is stored in a read-only memory (ROM) chip physically separated from the controller circuit until the weapon is to be armed for drop. This separation is intended to prevent accidental generation of an enabling pulse train in case of an accident. A single incorrect pulse will lock up the switch and prevent closure.

The MC2969 is also used with the W78/MK I 2A/Minuteman III weapon system. In this application the switch is enabled if the Minuteman missile delivers to the weapon prearm electronics a proper three-stage acceleration profile and a signal from the guidance system indicating proper guiding. This acceleration profile is a result of the three stages of the Minuteman, each of which burns its propellant to exhaustion and then falls away. The resulting acceleration profile ramps to acceleration peaks at the instant before propellant exhaustion. The acceleration then drops sharply to near zero as the expended stage falls away and thrust of the next stage begins but gradually increases during the burn time of the lower-thrust next stage. The resulting waveform is a three-peak acceleration-time sawtooth. This sawtooth and the generation of a good guidance signal are uniquely associated with proper performance of the Minuteman missile. This design is intended to provide high probability that peacetime accidents will not generate the requisite prearming inputs. Weapons containing this switch first entered the national stockpile in 1977. From 1977–1993, 852 switches (as parts of a weapon) were subjected to new-material drop or flight tests or drawn from the stockpile for examination and drop/flight testing. For the new-material and stockpile evaluations no switches have been found in the closed position, nor have they failed to close when properly signaled. However, 7 countable failures have occurred in 1,100 tests at the component and next assembly levels. The reliability assessment of the switch is 0.996. No nuclear tests are required to continue this assessment process.

The Device

The device is a special case. As noted above, it is sophisticated but not complicated. The device designers hold that its performance is guaranteed if it is manufactured according to exacting specifications prepared during the design phase of the program. During this phase, the device is carefully tested, e.g. with yield tests. During production and while in stockpile, it (as part of the weapon) undergoes new-material and stockpile evaluations without yield testing. In the recent past a few weapons have been drawn from the stockpile, modified in some cases to comply with the Threshold Test Ban Treaty (TTBT), and detonated underground. Because these tests were infrequent (a given weapon type underwent perhaps one test during its stockpile life), they do not assess reliability except in the grossest sense, i.e. either 1 or 0. Many observers view these tests as a demonstration that perhaps tests the designer more than the weapon. Such tests should be considered tests of confidence, not reliability.

Reliability Assessment of the Stockpile

With the introduction of completely assembled, maintenance-free weapons, the method of reliability assessment of the stockpile metamorphosed from total dependence on field-testing results to a stockpile sampling program. This program, begun in 1958, involved withdrawal of weapons and testing of the nondevice hardware at an initial rate of 50 weapons at 6-month intervals for the first 18 months of stockpile life. The sampling program could subsequently draw fewer weapons per cycle (perhaps 50 weapons per year), depending on initial evaluation results. All tests were laboratory examinations; no flight or drop test program was included.

The test program quickly revealed that most problems were related to design and/or production. As a result, a new-material evaluation program was introduced in 1959 that examined newly produced weapons immediately after assembly at the final assembly plants. In this program, a few weapons were examined each month at production outset. Based on initial findings, this rate was modified in later production. Again, only laboratory tests were performed.

In 1963 the new-material evaluation program was expanded to include joint AEC/DoD-sponsored flight and drop tests—generally four per year for most weapon types. Randomly selected weapons of a given type are withdrawn from the national stockpile and returned to the DOE final assembly plant (Pantex) near Amarillo, Texas. The device is separated from the remainder of the weapon, and an instrumentation/telemetry package is mounted in its place. The mass and dynamic properties of this package match as closely as practical the mass and dynamic characteristics of the device. This package monitors the performance of the safing, arming, fuzing, and firing system as well as inputs from the carrier vehicle. Where appropriate, sensors are included to monitor the dynamic performance of the test item. This pseudoweapon, called a Joint Test Assembly, is then returned to the military for installation in a delivery vehicle for a weapon system test in a realistic environment. The results are telemetered to one or more receiving stations for analysis and evaluation.

Beginning in 1970, accelerated aging units (AAUs) were added to the evaluation program in an attempt to obtain early warning about significant materials-related degradation. One or two complete weapons from early production were subjected to temperature cycling and high-temperature storage for one or more years. The interior volume was monitored by gas sampling. The AAUs were disassembled periodically for more complete inspection. AAU activity is not considered a formal part of the stockpile evaluation program because its environmental exposure is not representative of normal stockpile conditions. This activity is admittedly qualitative because there is no known way to predict an exact accelerated aging factor owing to the complexity of the reaction kinetics as a result of the numerous activation-energy coefficients associated with the materials in the weapon. Nevertheless, the information obtained from the accelerated aging program is important to the design and stockpile-evaluation community in its attempt to forecast aging effects.

In the mid-1980s the evaluation program was again rebalanced to further emphasize the new-material assessment based on expanding data bases, which continued to indicate that most defectiveness resulted from design or production errors, not from degradation. The stockpile-evaluation portion of the program was relaxed to biennial sampling of 11 weapons per cycle after completion of production, another indication that the weapon types in the stockpile, when well designed and produced, exhibit good age stability. For example, some weapon types have been in the national stockpile for 25–30 years and with periodic attention exhibit no significant reliability degradation.

Since the start of the current stockpile evaluation and reliability assessment program in 1958, ~13,000 weapon evaluations have been conducted. During this period, the failure rate of the nondevice hardware suggests an expected weapon failure rate of 1–2% for the stockpile. This forecast assumes that the device has a reliability of one. The nondevice hardware is not perfect, and retrofits have been necessary to correct the problems noted above, the majority of which have been caused by design oversights and/or fabrication/assembly mistakes. Military service mishandling and aging-induced defectiveness represent only a small fraction of the total unreliability. Twice a year the DOE issues a document containing one or more reliability assessments for each weapon type in the stockpile. Several assessments are sometimes necessary to address different deployment options.

Given that the stockpiled weapon types underwent thorough and complete design, development, testing, and evaluation, we have no compelling arguments for the necessity of continuing yield testing to retain confidence in the reliability of hardware for an extended period of time. To maximize the useful life, one can take such steps as refrigerated storage, limited movement, no modifications, aggressive surveillance, etc. Most importantly, in the case of a test ban, one should not tamper with the device hardware once it has been certified. An article by Fletcher Pratt[18] occasionally has been cited as an argument for continued yield testing. We believe that, instead, the situation Pratt described supports the argument that thorough design, development, test, and evaluation are required before deployment, and in the absence of follow-on stockpile testing, hardware modifications must be avoided. The torpedo failures resulted from violations of all of these steps. New, untested devices should not be considered.

Notes

1. Robert Peurifoy, "A Personal Account of Steps Toward Achieving Safer Nuclear Weapons in the U.S. Arsenal," chapter 2 in this volume.

2. Samuel Glasstone and Philip J. Dolan, eds., *The Effects of Nuclear Weapons*, 3rd ed. (Washington, D.C.: U.S. Government Printing Office, 1977).

3. The standards are stated in full in Sidney D. Drell, John S. Foster Jr., and Charles H. Townes, "The report of the nuclear weapons safety panel," hearing before the Committee on Armed Services, House of Representatives, 101st Cong., 2nd sess., December 18, 1990, vol. 5 (Washington, D.C.: U.S. Government Printing Office, 1990).

4. Christopher H. Stubbs, "The Interplay Between Civilian and Military Nuclear Risk Assessment, and Sobering Lessons From Fukushima and the Space Shuttle," chapter 3 in this volume.

5. Marvin L. Adams and Sidney D. Drell, "Technical Issues in Keeping the Nuclear Stockpile Safe, Secure, and Reliable," and Bruce Goodwin and Glenn Mara, "Stewarding a Reduced Stockpile," LLNL-CONF-403041, April 2008. See "Nuclear Weapons in 21st Century U.S. National Security," Supporting Papers, 11–25 (Center for Strategic and International Studies, American Physical Society, and American Association for the Advancement of Science, 2009).

6. A detailed discussion of these safety and reliability issues and technology is given in the appendix to this paper in sections 3 and 4, excerpted from a paper by Sidney Drell and Robert Peurifoy entitled "Technical Issues of a Nuclear Test Ban," published in *Annual Review of Nuclear and Particle Science*, 1994.

7. Ibid.

8. However, the geometric configuration of warheads in the Trident I and II systems has called this assumption into question. Later in this section this issue is discussed further. The discussion of this section on safety draws on Drell et al., "Nuclear Weapons Safety Panel House Armed Services Committee."

9. "Normal environments are those expected logistical and operational environments, as defined in the weapon's stockpile-to-target [STS] sequence and military characteristics in which the weapon is required to survive without degradation in operational reliability."

10. "Abnormal environments are those environments as defined in the weapon's stockpile-to-target sequence and military characteristics in which the weapon is not expected to retain full operational reliability."

11. It is estimated that only 0.003% of the larger ^{239}Pu particulates created in a deflagration are retained in the gastrointestinal system, in contrast to ~3.6% of inhalation-ingested Pu. Moreover, no more than ~0.2% of exposed Pu is released in respirable form by fuel-fed (1000°C) fires, in contrast to ~10–20% in detonations.

12. E. James, *Report UCRL-LR-113578*, Lawrence Livermore National Laboratory, 1993.

13. Drell et al., "Nuclear Weapons Safety Panel House Armed Services Committee."

14. John R. Harvey and Stefan Michalowski, *Center for International Security and Arms Control*, Stanford University, 1993.

15. Drell et al., "Nuclear Weapons Safety Panel House Armed Services Committee."

16. Harvey and Michalowski, *Center for International Security and Arms Control.*

17. This switch is not a use-control feature. The pulse train required to change the state of the switch is a unique pattern, not a classified code. Use-control needs are accomplished by other methods, e.g. two-man rules, security forces, and permissive action links (PALs). More recent detonation-safety themes rely on "other-than-electrical" power and signal control, e.g. magnetic and optical coupling, as used in the W88 Trident II warhead and described in K. G. McCaughey, In *Characteristics and Development for the MC3831 Dual Strong Link Assembly, Sand88-1412.6*, Sandia National Laboratories.

18. Fletcher Pratt, *Atlantic Monthly* 186:25 (1950).

A Personal Account of Steps Toward Achieving Safer Nuclear Weapons in the U.S. Arsenal

2

ROBERT L. PEURIFOY

I spent my professional career at Sandia National Laboratories, working for thirty-nine years on nuclear weapons. This paper was prepared in response to Professor Sidney Drell's request to discuss the history of developments to achieve a high level of nuclear weapon safety.

The field of play for establishing nuclear weapon design foci is illustrated in Figure 2.1. There are a number of players in this game, not all with the same points of view. Quadrants 1, 3, and 4 are for the most part noncontroversial; quadrant 2 is contentious. For war fighters:

- More safety means less reliability as a result of the additional operations that are introduced to prevent an accidental detonation.
- More security means higher costs for the Department of Defense (DoD) and the Department of Energy (DOE).
- More control means less immediate availability as a result of additional operations that are introduced to prevent an unauthorized detonation.

Sid Drell[1] displayed a list of thirty-two "Broken Arrows" (accidental events leading to losses, disappearances or crashes involving nuclear weapons) in the paper which is now an appendix to chapter 1 of this volume. Consider number twenty-one. On January 24, 1961, a B-52 broke

Figure 2.1 Nuclear Weapon Design Foci

	PEACETIME EMPHASIS	WARTIME EMPHASIS
IMPROVE	Safety Security Control	Survivability Deliverability Effectiveness Flexibility Battle Management
REDUCE	Maintenance Movement Training	Reaction Time Operational Constraints Collateral Damage

apart in flight near Goldsboro, North Carolina. The B-52 was involved in the Strategic Air Command (SAC) Airborne Alert program (Operation Chrome Dome), which involved flying B-52s carrying nuclear weapons. During the breakup, two B39 four-megaton bombs were released from the aircraft. One got its picture in the paper (Figure 2.2). This bomb did exactly what it was designed to do. A sequence of steps was activated as it fell to earth and prepared it to explode exactly as it was designed to. Fortunately, the one arming control switch located in the cockpit also did what it was designed to do. It remained in the *safe* position because it had not been moved to the *arm* position by the pilot.

Keep this fact in mind: sometimes alert-configured aircraft suffer electrical faults. In the absence of enhanced nuclear detonation safety (ENDS), with two strong links as discussed by Drell in his chapter and more fully in the appendix to his paper, if a specific wiring fault had occurred during breakup, the bomb could have exploded as designed. Fortunately, in this case a short in the wiring did not bypass the pilot's authority.

Again, using the Broken Arrow list (accident summary), I call your attention to Broken Arrow number twenty-nine (Palomares, Spain), January 17, 1966, and number thirty (Thule Air Base, Greenland), January 21, 1968. In the Palomares accident, as with the Goldsboro accident, the bombs separated from the B-52 as it broke apart in the air. The danger of a nuclear detonation was similar to that in the Goldsboro accident,

Figure 2.2 A four-megaton thermonuclear bomb (B39) hanging in the branches of a tree. It was dropped during the break-up of a SAC bomber (B-52) on airborne alert over Goldsboro, N.C., in 1961. Its parachute opened and it landed with minimal damage. Fortunately one of the switches in the arming sequence remained in the safe position.

and it was avoided for the same reasons: the pilot did not move the arming control switch to the arm condition and there were no "arming circuit" shorts in the wiring. In the Palomares accident, the B-52 carried four bombs. At ground impact, the chemical high explosive detonated in two of the bombs, dispersing plutonium. The bombs met the one-point abnormal environment safety criterion and the detonations did not initiate a fission chain reaction leading to a nuclear yield.

In the Thule accident, the bombs did not separate from the B-52 as it crashed. Therefore, the bombs' safety components did not experience all of the bomb free-fall environments required to arm them. However, at ground impact the chemical high explosive in the bombs did detonate, and plutonium was scattered.

Nothing was done to improve safety by retrofitting the bombs already deployed in the U.S. arsenal.

In 1968, as a result of the Palomares and Thule incidents, the SAC terminated Operation Chrome Dome and Sandia established a central, focused safety group headed by Bill Stevens, a second-level supervisor. Prior to 1968, each weapon design group chose its own detonation safety architecture. The risks associated with abnormal (accident) environments were not well recognized by Sandia senior management until 1968. Before 1968, priority was given to reliability rather than safety. (There was a central, focused reliability group at Sandia starting in the early 1950s.) By 1971, Stevens's group had formulated ENDS technology.

Stimulated by Sandia studies (Project Crescent, 1971) of brute-force engineering methods to reduce the risk of plutonium-scattering in an accident, the Los Alamos National Laboratory (LANL) investigated the use of insensitive high explosives (IHE) for air-delivered weapons. The first IHE yield test was conducted by LANL in 1974.

In the fall of 1973, I was put in charge of the Sandia/Albuquerque Weapon Design Directorate. I quickly did two things:

- I required that all new weapon designs use ENDS technology.
- I recommended to my vice president, Glenn Fowler, that we push for a safety upgrade of weapons then in the stockpile. Glenn enthu-

siastically agreed and arranged a briefing for Sandia senior management in February 1974.

The second recommendation was not well received by the Sandia senior management.

In April 1974, Fowler and I took the retrofit recommendation briefing to the Director of Military Applications of the Atomic Energy Commission. He and his staff were passive. Fowler felt we should go on written record about our concerns. He asked me to have a letter prepared for his signature detailing our concerns and addressed to the Director of Military Applications. He signed the letter on November 15, 1974. A redacted copy of his letter is contained in Appendix 2.1 (see pages 75–80). Fireworks erupted in Washington, because plausible deniability had been destroyed.

In 1976, fifteen years after Goldsboro, ten years after Palomares, and eight years after Thule, the first bombs designed with ENDS, the new B61, entered into the stockpile.

In 1979, versions of the B61 were the first bombs to enter the stockpile with both ENDS and IHE. In 1981 the first missile warhead entered the stockpile with both ENDS and IHE. B-52s were loaded with W80 warheads on air-launched cruise missiles.

On September 15, 1980, at the Grand Forks, North Dakota, Air Force Base, an engine pod on an alert-configured B-52 caught fire while undergoing engine-start for a take-off drill. Fortunately, a steady wind kept the flames confined to the engine pod and away from the B-52 fuel tanks for three hours. It is puzzling that this incident was not declared a Broken Arrow. The weapons on board did not contain ENDS or insensitive high explosives.

Another incident occurred August 19, 1980, at Robins AFB in Georgia, when a B-52 was destroyed by a ground fire. In spite of occasional claims that B-52s don't catch fire, they do.

I continued to press my concerns about the existing risks of high-consequence safety failures in the deployed nuclear weapons but to no avail until, in 1988, the acting deputy assistant secretary for military application in the Department of Energy, John Meinhardt, contracted with Pacific Sierra Research to conduct an independent weapon safety study. The study was led by Gordon Moe,[2] a Pacific Sierra vice president, and supported my concerns as expressed in this concluding thought:

> Attention to safety has waned, and we still have risks from weapons that will remain in the stockpile for years. The potential for a nuclear weapon accident will remain unacceptably high until the issues that have been raised are resolved. It would be hard to overstate the consequences that a serious accident could have for national security.

The Moe report had little, if any, impact prior to April 26, 1989, when Senator John Glenn (D-Ohio) visited Sandia. I gave him a thirty-minute safety briefing, using a picture of the Grand Forks fire, a display of the many safety briefings given to government officials (about 800), and the Moe study. The senator asked me what Admiral Watkins (then Secretary of Energy James D. Watkins) thought of my safety concerns. I said that Admiral Watkins did not know of my concerns. Senator Glenn then said, "I'll be traveling with Admiral Watkins to the Savannah River site next week, and I'll discuss this with him." Fireworks erupted among Sandia and DOE senior management. Admiral Watkins was immediately informed and was subsequently briefed by Gordon Moe, with Secretary of Defense Dick Cheney in attendance. This led to an immediate decision by Secretary Cheney to remove the SRAM from alert status and caused the assistant to the secretary of defense for atomic energy (ATSD-AE), Robert Barker, to initiate a quantitative safety study.

Soon thereafter, the House and Senate armed services committees learned of these safety concerns. They held hearings in May 1990 in which Watkins and the lab directors testified. Much dodging took place.

Eventually, a letter dated May 1990 was sent to Watkins from Chairman Les Aspin and Ranking Member William Dickinson of the U.S. House of Representatives Committee on Armed Services; John Spratt and Jon Kyl, respectively chairman and ranking member of the committee's panel on DOE's nuclear facilities; with the endorsement of the leaders of the Senate Armed Services Committee. The letter notified the secretary of energy that "we have empaneled three eminent physicists, Dr. Sidney D. Drell of Stanford University, Chairman, Dr. John S. Foster Jr. of TRW Corporation, and Dr. Charles H. Townes of the University of California at Berkeley to evaluate these issues and to provide us with their advice." The issues involving the safety of the U.S. nuclear forces had been raised during a recent hearing before the Department of Energy Defense Nuclear Facilities Panel. The letter requested Watkins's assistance.

On May 23, 1990, the *Washington Post* published an article written by R. Jeffrey Smith dealing with safety concerns involving three nuclear weapon types (see Appendix 2.2 on pages 81–88). The W69/Short Range Attack Missile (SRAM) was one of them.

On December 9, 1990, as a result of the Smith article, Secretary of Defense Cheney instructed the Air Force to permanently remove all nuclear weapons from B-52 "quick reaction alert" status.

The Drell Panel Report was released[3] in December 1990.

With the case for nuclear weapon safety bolstered by outside support and public exposure in the media, as of August 2010 all nuclear weapons in the active stockpile contain ENDS technology. All air-delivered weapons contain IHE. Table 2.1 summarizes the many steps, spread out over forty-two years, to achieve this goal. The timing intervals show convincingly how important outside support and public exposure were in achieving a high level of nuclear weapon safety.

Table 2.1 Nuclear Weapon Safety Evolution

1961	Goldsboro, North Carolina
1966	Palomares, Spain
1968	Thule Air Base, Greenland
1968	Sandia formed a central, focused Nuclear Safety Department with Bill Stevens named manager.
1968–1971	ENDS architecture formulated.
1973	Peurifoy put in charge of Sandia Albuquerque weapon program.
1973	ENDS introduced into weapon designs.
1973	Old bomb retrofits proposed by Fowler/Peurifoy but rejected by Sandia management.
1974	Fowler letter. Much anger, AEC, DoD, Sandia.
1976	ENDS enters stockpile, B61.
1980	Grand Forks, N.D., incident
1988	Gordon Moe study
1989	Senator John Glenn visits Sandia. Peurifoy gives safety briefing.
1989	Admiral Watkins is made aware of safety concerns.
1989	ATSD-AE demands a quantitative safety study. Study supports Peurifoy concerns.
1990	House and Senate committees learn of safety concerns.
1990	R. Jeffrey Smith, *Washington Post*, breaks safety story.
1990	The Drell Panel
1990	Secretary of Defense Cheney orders removal of nuclear weapons from B-52 QRA force.
1991	Peurifoy takes early retirement.

Appendix 2.1: Sandia VP Fowler Notification of Safety Concerns Addressed to Major General Ernest Graves, U.S. Atomic Energy Commission

UNCLASSIFIED (excerpt)

Secret

SAND85–0474 •
Nuclear Weapon Data • Sigma 1
Specified External Distribution Only
~~Printed April 1986~~

RS3140/92/00001
THIS DOCUMENT CONSISTS OF 40 PAGE(S)
NO. 1 OF 1 COPIES, SERIES A

(1-15-92)

Final Development Report for the B61-7 Bomb (U)

Redacted Copy

SAA200034780000

Sandia National Laboratories, Los Alamos National Laboratory

Prepared by
Sandia National Laboratories
Albuquerque, New Mexico 87185 and Livermore, California 94550
for the United States Department of Energy
under Contract DE-AC04-76DP00789

INVENTORIED
DATE

Classified by D. L. McCoy, Supervisor, B61/Stockpile Improvement Division, 5111, February 22, 1985.

CRITICAL NUCLEAR WEAPON DESIGN INFORMATION
DOD DIRECTIVE 5210.2 APPLIES

~~RESTRICTED DATA This document contains Restricted Data as defined in the Atomic Energy Act of 1954. Unauthorized disclosure subject to Administrative and Criminal Sanctions.~~

DEPARTMENT OF ENERGY DECLASSIFICATION REVIEW	
1ST REVIEW-DATE:	DETERMINATION [CIRCLE NUMBER(S)]
AUTHORITY: ☐AOC ☐ADC ☐ADD	1. CLASSIFICATION RETAINED
NAME: FIELD (ALBO)	2. CLASSIFICATION CHANGED TO:
2ND REVIEW-DATE: 8/24/98	3. CONTAINS NO DOE CLASSIFIED INFO
AUTHORITY: ADD	4. COORDINATE WITH: DOD COORD. COMPLETE
NAME: M. L. Kolbay	5. CLASSIFICATION CANCELLED
	6. CLASSIFIED INFO BRACKETED
	7. OTHER (SPECIFY): SECOND REVIEW ONLY

EXTRACT COPY - APPENDIX A ONLY

NOTICE Reproduction of this document requires the written consent of the originator, his successor or higher authority.

***Only those recipients external to SNL as listed under "Distribution" are authorized to receive copies of this report. They are not authorized to further disseminate the information without permission from the originator.**

I9 2-3219 1/1A $

RS 3140/92/00001, Series A, 1 Copy, 40 Pages, SRD, dtd 1/15/92
Subject: Final Development Report for the B61-7 Bomb (U)

Distribution: 1/1 3180 – R. B. Craner

SERIES B: This is an exact copy: 3180: tcb: 3/31/92
1 - 3180 R.B. CRANER

Sandia National Laboratories   Los Alamos National Laboratory

RECEIVED
PC/DAS
3180

UNCLASSIFIED

Secret

APPENDIX A

Program Documents

RS 3140/92/00001

~~SECRET~~ UNCLASSIFIED

COPY

COPY

Sandia Laboratories

Albuquerque, New Mexico 87115

G. A. Fowler
Vice President, Systems

Major General Ernest Graves
Assistant General Manager for
Military Application
Division of Military Application
U.S. Atomic Energy Commission
Washington, D.C. 20545

Subject: Safety of Aircraft Delivered Nuclear Weapons Now in Stockpile

Ref:
1. Uncl. letter, Major General Ernest Graves, DMA to G. A. Fowler, dtd 9/12/74
2. SRD, Stockpile-to-Target Sequence for the B61 Bomb, RS 3141NC/501382, dtd 15 February 1974
3. SRD, Military Characteristics for the Warhead for the Trident Mk4 Re-Entry Body (U), RS 3148-1/101274, dtd 30 August 1973
4. SRD, B77 MC's (Military Characteristics for a New FUFO Bomb (B77)) (U), RS 3141NC/501741. dtd 14 August 1974
5. SRD, "Project Crescent: A Study of Salient Features for an Airborne Alert (Supersafe) Bomb (U), SC-WD-70-879, RS 3410/2097, dtd April 1971
6. SRD letter, D. P. MacDougall, LASL and G. A. Fowler, SLA to Major General Ernest Graves, DMA. ADW-477, RS 3148-1/102306, dtd 2/15/74, and enclosure, B61-3 and 4 Safing/Denial Study, ADW-PM-74-53, RS 3148-1/102307, dtd 2/25/74

Most of the aircraft delivered nuclear weapons now in stockpile were designed to requirements which envisioned weapon stockpile operations consisting mostly of long periods of igloo storage and some brief exposure to transportation environments. Changing conditions in the early 1960's dictated different operational practices which included wide spread ground and air alert operations. Starting in 1968 new weapon STS's have gradually accounted for this change in weapon usage by providing more realistic abnormal environment definitions. Reference 1 acknowledges this trend toward recognizing realistic abnormal environments and suggests that future MC's, in consonance with this trend, should require that nuclear weapons be designed to meet current safety requirements in the presence of fault signals applied to the weapon. This philosophy is consistent with the wording of the new B61 STS (calls out fault signals as an abnormal environment), the W76 MC's and the B77 MC's (Ref. 2, 3, and 4). We agree with the validity of this approach and in the case of aircraft delivered weapons believe the need for this policy is well demonstrated by the many prearming incidents involving direct current driven Ready/Safe Switches (summarized in Attachment 1). Both the B61-3 and 4 and the B77 are being designed to meet these new requirements so long as the unique signal override feature remains in the NORMAL position which requires that the weapon receive a unique prearming signal and therefore cannot be inadvertently prearmed by any other power source in the aircraft or in handling or test equipment.

In 1968, Sandia Laboratories established a safety assurance program to study and understand the implications of designing nuclear weapons for safety in the abnormal environments. Reference 5 reports the results of a study commissioned by DMA related to aircraft/weapon safety. A product of this effort was the conception of the strong link/weak link/exclusion region principle on which the new safety technology is based. A study of the abnormal environment safety of stockpile systems was initiated in 1970 and was intensified late last year with priority given to aircraft delivered systems because of the frequency of Ready/Safe Switch incidents and the history of aircraft related accidents involving nuclear weapons.

Interim results from the priority portion of this review are now available and are provided along with Sandia Laboratories' conclusions and recommendations.

The following is a compilation of weapon safety requirements at the time of stockpile entry (Table I) and a brief description of each weapon safing scheme along with our conclusion regarding the adequacy of safety in the abnormal environments.

COPY COPY

TABLE I

System	Ø 3	Ø 6	Normal Env.	Abnormal Env.
W25 (GENIE)	11/54	1/57		
W28 (HOUND DOG)	8/54	8/58		
B28 EX & RE	8/54	8/58		
B28 FI	9/60	7/62		
B43	10/56	2/60		
B53	12/58	8/62		
B57	1/60	1/63		
B61-0, 1, 2	1/63	1/68		
W69 (SRAM)	1/67	2/72		
W72 (WALLEYE)	5/69	9/70		

DoD (b)(3)

Note 1: Fire only abnormal environment specified.
Note 2: In the absence of input signals except normal monitor and control.
Note 3: Fire and shock only abnormal environments specified.
Note 4: Fire, shock, F-4 aircraft crash, fragmentation, nuclear radiation, lightning, and flooding specified as abnormal environments.

The W25 and W69 warhead each contain a single environmental sensing safety feature, an integrating accelerometer. The W28 warhead contains no environmental sensing safety feature. It does contain a 28 volt DC motor-driven high-voltage safing switch, controlled by aircraft power. The W72 warhead contains a single environmental sensing safety feature, a velocity-sensing differential pressure switch.

Each of the bombs contains one active environmental sensing safety feature for each option; integrating accelerometers, velocity-sensing differential pressure switches, or hydrostats (B57, ASW). In addition, they each contain one or two 28 volt DC motor-driven safing switches, controlled by aircraft power. None of these safety switches (Ready/Safe and Environmental Sensing), with the exception of the W69 ESD, have any hardening features which would help to assure safety during exposure to abnormal environments. All of the 28 volt DC safing switches will arm if supplied with typical aircraft stray voltages and currents through fault circuits in the aircraft as specified in the B61-3, 4 STS (Ref. 2).

(b)(3)

In summary, all of the current stockpile of aircraft delivered weapons (and the B61-2 entering stockpile this fiscal year) have serious shortcomings when evaluated against current abnormal environment nuclear safety standards. These shortcomings stem from the inability of existing safing devices to assure the maintenance of a predictably safe state through exposure to abnormal environments, the possibility of these safing devices being electrically bypassed through charred organic plastics or melted solder and finally the susceptibility of the safing devices themselves to premature operation from stray voltages and currents which may be present in the abnormal environments.

COPY COPY

RS 3140/92/00001

~~SECRET~~ UNCLASSIFIED

It appears that the safety of the aircraft delivered stockpile could be greatly improved over the next decade in the following manner:

1. Retire the following weapons or retrofit them with two independent safety devices utilizing the strong link/weak link concept:

   W25 (GENIE)
   W28 (HOUND DOG)
   B57 (ASW)
   B53
   B61-0, 1, 2
   W69 (SRAM)
   W72 (WALLEYE)

2. Replace the following weapons as indicated:

   B28 EX/RE – Replace with B61-3,4,5 and B77
   B28 FI – Replace with B77
   B43 – Replace with B61-3,4,5 and B77
   B57 (TAC) – Replace with new FUFO MRR and/or NATO bomb

As you pointed out in our conversation earlier this month, a plan to modernize or replace the aircraft-delivered weapons to improve safety is a subset of a broader stockpile modernization and retirement plan. Perhaps the urgency associated with the safety question will serve to stimulate the effort associated with the overall plan. We will be glad to help in any way we can either with the abnormal environment safety plan or with the broader question.

/s/Glenn A. Fowler

CCB:1511:ps

RS 1000/4465, Series A, SRD, 6 pages, 12 copies

Distribution:

1 - M1382 Maj. Gen. Graves, DMA
2 - M0828 R. E. Batzel, LLL
3 - M0828 H. L. Reynolds, LLL
4 - M0801 T. B. Cook, Jr., 8000
5 - M0737 H. M. Agnew, LASL
6 - M0737 E. H. Eyster, LASL
7 - M0659 H. C. Donnelly, ALO
8 - M0659 W. R. Cooper, ALO, Office of Plans & Budgets
9 - M0659 J. F. Burke, ALO
10 - M0659 T. C. Jones, ALO
11 - 1 Morgan Sparks
12 - 1000 G. A. Fowler

COPY

COPY

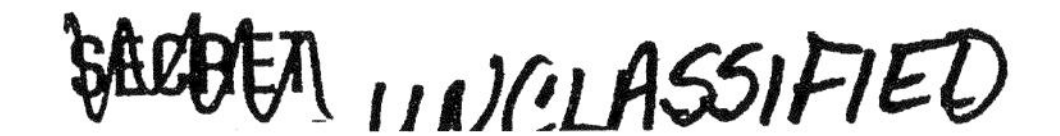

RS 3140/92/00001

COPY

COPY

RS1000/4465, Series B, SRD, 6 pages, 7 copies, November 15, 1974

Series B Distribution: Series 7230-C - THIS IS AN EXACT COPY:2/12/85:

1 - 5100 R. N. Brodie

1 - M0801 L. Gutierrez, 8100
2 - 1500 R. L. Peurifoy, Jr.
3 - 1520 G. J. Hildebrandt
4 - 1600 H. E. Lenander
5 - 1650 W. L. Stevens
6 - 2000 K. D. Bowers
7 - 9500 L. J. Heilman

COPY

UNCLASSIFIED

SECRET

COPY

Appendix 2.2: Defective Nuclear Shells Raise Safety Concerns; U.S. Secretly Repairing Weapons in Europe

***Washington Post*, May 23, 1990**

By R. Jeffrey Smith

In a series of secret moves, the U.S. government discovered—and is repairing—defective nuclear artillery shells that could have exploded accidentally while stored in Europe, and has started urgent studies of the designs of two other atomic warheads whose safety is suspect, according to senior U.S. officials and weapons scientists.

In the most serious incident, the government in 1988 belatedly discovered a defect in the W79 short-range artillery shell after it had been deployed in Europe. Urgent orders were issued not to move the warheads, and repair teams hurried to the nuclear ammunition depots to disable the several hundred shells so they could not be accidentally detonated, the officials and scientists said.

The artillery shells, which are deployed in at least three West European countries and in U.S. stockpiles, are being modified so they will detonate only after being fired in battle, the officials said.

The Joint Chiefs of Staff, in a separate nuclear weapons safety incident, last year secretly imposed special restrictions on the handling and deployment of a short-range missile carried by U.S. strategic aircraft in order to avoid accidental explosions that could disperse cancer-causing, radioactive plutonium from the missile warhead, these sources said.

The Department of Energy, in a third case, last week agreed to a secret congressional request for an independent scientific inquiry into the possibility that a nuclear warhead now being deployed aboard Trident strategic submarines could be detonated in a possible missile-handling accident in port, according to the sources.

The problems with the three weapons have raised serious questions on Capitol Hill and in the Bush administration about U.S. nuclear weapons safety. Some experts forecast that additional problems may be uncovered by special scientific inquiries at the nuclear weapons laboratories ordered recently by Secretary of Energy James D. Watkins.

The Department of Energy (DOE) is responsible for the design and production of all nuclear warheads. The Department of Defense (DoD) determines warhead requirements and develops the weapons on which the warheads are deployed.

Watkins, while declining to discuss difficulties with specific weapons, said in an interview this week that "we're not all that comfortable" with the government's past approach to several safety issues, and that "we need to focus a lot more" on measures to diminish the risk of accidentally dispersing plutonium in warheads.

Watkins also emphasized that when weapons safety problems are uncovered, "we take the operational steps necessary to minimize those risks, take the weapons out of service if necessary and then look for some mid-term and long-term fixes. We're in that process right now."

He added in response to questions that "fortunately, we don't have many {weapons} that fall into that category, but I can tell you that we're serious about those things and we do what's necessary to make sure that we don't have any situation . . . where we can't meet our [safety] specifications."

In contrast to the wide attention recently given to environmental and scientific problems of nuclear weapons production, safety questions about individual weapons have scarcely been scrutinized outside the tight-knit community of weapons designers and the defense officials to whom they answer.

Some high-level Bush administration officials say they believe that safety issues traditionally have had a much lower priority than military concerns such as increasing a nuclear weapon's explosive power, efficiency or reliability.

But Watkins, who says he came into office in 1989 with a "very significant sense of laxity in the safety practices of DOE," has ordered what other officials say is the first comprehensive assessment of the probability of an accident involving any of the more than 20,000 nuclear warheads deployed with U.S. forces around the globe.

No nuclear weapon ever has been known to detonate accidentally and produce a nuclear yield. However, there have been unexpected detonations of the volatile chemical explosives surrounding the nuclear materials in warheads, including several incidents that resulted in considerable radioactive contamination.

Officials say that Watkins's safety concern about the Short-Range Attack Missile-A, or SRAM-A, carried by strategic bombers led to establishment of a special committee on nuclear weapons safety last year that includes senior DoD and DOE officials.

Although it will report simultaneously to Watkins and Defense Secretary Richard B. Cheney, the two departments, in a conflict indicating diverging priorities on weapons issues, fought a heated battle over who would chair the committee. The dispute eventually was won by DOE.

"I felt it was necessary that we have a somewhat independent committee . . . that could stand off from military requirements, military demands and focus heavily on the safety issue," Watkins said. "I just felt that conflict of interest ought to be separated out."

The information about recent nuclear weapons safety problems in this article is derived from interviews with more than two dozen U.S. military and civilian officials and nuclear weapons scientists. None was willing to be quoted by name, because everything about the episodes is highly classified.

Officials say the most dangerous and politically sensitive incident was the surprise discovery in early 1988 that W79 artillery shells deployed in West Germany, Italy and the Netherlands could accidentally explode if they were struck forcefully at a sensitive spot, perhaps by a stray bullet or impact from a nearby battlefield explosion in wartime.

A single, elliptical reference to the problem appeared in an unclassified report issued by Watkins three months ago. The report mentioned that a warhead developed by Lawrence Livermore National Laboratory under DOE supervision, identified only as "WXX," had recently caused "one-point safety concerns" that were confirmed by underground nuclear testing.

U.S. officials subsequently confirmed that the "WXX" was the W79.

"Nobody was blase about it, but nobody was panicky either," said a senior military official of the government's reaction. "We did not foresee an imminent catastrophe. But when it comes to nuclear safety, we treat everything as potentially serious."

Computer calculations and underground tests before the start of production in 1981 had indicated no safety problems. But a new safety analysis at Livermore in 1988 raised concerns that were confirmed by secret underground nuclear tests in December 1988 and February 1989, officials said.

The tests indicated that the W79 did not meet a secret government safety standard requiring that with any warhead design, under any circumstances, there be less than a one-in-a-million chance of an accidental nuclear explosion with a yield as powerful as a blast from about four pounds of TNT, enough to destroy a small room.

Several officials said an inadvertent explosion of the material surrounding the nuclear core was particularly likely to produce a nuclear yield if it occurred while the shells were loaded inside the 8-inch howitzers from which they are fired, an unusual circumstance in peacetime. But a senior military official said, "For a while, we were also worried that these things might go off if they fell off the back of a truck and landed in a certain way."

The officials did not say how big an accidental explosion a W79 shell might have caused. The warhead is designed to explode in battle with up to a 10-kiloton nuclear yield, about two-thirds the force of the 1945 Hiroshima bomb.

A highly-placed foreign official said that after the confirming nuclear test, a small group in the West German government was told in a general way that "there was a chance of technical failure leading to an explosion" of the shells and that "some adjustments" to their design were needed to prevent any accident.

Senior West German officials "were not informed explicitly" how an accidental nuclear blast might occur, the official said. "But I do not rule out that a more detailed briefing was given to specialists" in the German ministry of defense, he added.

Officials said the information was kept otherwise secret to avoid panicking citizens or calling into question the viability of the U.S. nuclear force in Europe, which includes dozens of nuclear-tipped missiles, more than 1,000 nuclear bombs and hundreds of older, nuclear-tipped artillery shells.

"It was obviously a politically hot potato," a U.S. official said.

Another senior military official said the episode alarmed the Pentagon and induced tensions with DOE. "It was the sort of problem that never should have occurred," the official said. "There simply was no good excuse for it."

After ordering that all W79 warheads be immobilized at their storage sites, the government sent teams of experts in early 1989 to install special "safing mechanisms" to block any detonation of the shells, several officials said.

Some of the warheads have been returned to the Pantex warhead production plant in Amarillo, Tex., so the "safing mechanisms" can be disabled and additional steel plating installed inside the skin of the shells at particularly sensitive spots.

Of the W79 problem, Watkins said only, "Without any question, safety has been preserved." Pentagon spokesman Pete Williams said last night that "the point is: the weapon is safe." He added that "changes were made to the W79, but I can't discuss the details of that."

The shells are in special U.S. storage bunkers overseas and at the Seneca Army weapons depot in New York state, including some that are still inoperative. Asked why, a senior official said, "it has to do with politics within the [Western] alliance."

But the Nuclear Weapons Council, three senior DOE and DoD officials who decide nuclear warhead production matters, was sufficiently concerned about the W79 to decide early this year that its design would not be replicated in a slightly smaller artillery shell, the W82.

"If one of these shells had a problem, then by definition the other one would certainly have it, too, in terms of the basic physics," one official said. "They basically have the same 'primary'," or nuclear core, he explained.

The decision forced at least a two-year suspension of W82 production, which had been scheduled to begin last February. Several sources said the delay was partly at the urging of Congress, which voted secretly last year to block W82 spending until the Bush administration certified that it was safe.

President Bush announced on May 3 that he wanted to halt the W82 deployment program in response to the declining military threat in

Eastern Europe. A senior U.S. military official said the design defects, which Bush did not mention, played no role in the decision.

Discovery of a safety defect in the W79 artillery shell prompted a more extensive review by the weapons laboratories of other warheads, which soon cast a shadow over the thermonuclear weapon now being deployed atop D-5 missiles in Trident submarines, the W88.

Some U.S. weapons scientists have alleged, based on computer modeling of accident scenarios, that the W88 could be detonated accidentally if the propellant fuel in D-5 missiles catches fire as a result of mishandling during loading operations at the Trident bases in Bangor, Wash., and Kings Bay, Ga.

A powerful nuclear blast or widespread dispersal of cancer-causing plutonium dust would result, these scientists say. They add that, in years past, the latter possibility has been taken so seriously that Livermore experts prepared maps of potential plutonium fallout over Spokane, Wash., near the Trident base at Bangor.

The allegations are the subject of a bitter scientific dispute at the highest levels of the Pentagon and DOE, according to some of the officials involved. The stakes are enormous, because the Trident missile system is expected to be at the heart of America's strategic deterrent force for the next three decades.

Although senior DOE and DoD officials say the risks are small, Watkins last week agreed to a secret, bipartisan congressional request that the issue be adjudicated by a special panel of three independent scientists cleared to review the nation's most sensitive nuclear weapons information.

Watkins said, "I have viewed all of the analysis, time and time again" on the W88, as have the directors of the three U.S. nuclear weapons laboratories, and "I'm satisfied . . . that we can continue to do the analysis we have to do on that weapon without undue concern." He said it now meets all nuclear explosive and weapons system safety standards.

At the same time, Watkins said that "had I been intimately involved in this process" during key deliberations in the early 1980s, "I would not have" made the decision to use the warhead's current design. "I don't think that kind of decision will ever be made again, and certainly won't be made while I'm here, and I believe with the kind of discussions that we've had with DoD it's not going to be made again."

At issue is the use of volatile explosive materials in the W88 warhead that scientists say would explode in a missile fire, producing forces that could compress the nuclear core in each bomb and begin a nuclear chain reaction. The Trident missile is considered particularly vulnerable to such an accident because its multiple warheads are arranged in a circle around the propellant fuel in the missile's third stage.

The warheads on most other U.S. ballistic missiles are arrayed on a platform that sits atop the final stage, allowing for the use of some form of insulating material to protect them from a missile fire.

Scientists at Livermore strongly protested the decision to use the volatile materials, but W88 designers at Los Alamos National Laboratory said that using a less volatile material would not have substantially diminished the risk of an accident. The Navy also opposed the idea because the added weight of the alternate materials would have reduced the missiles' range or required the deployment of fewer warheads on each missile.

Watkins said he would have "accepted the very modest penalties" associated with using the less volatile material. He said "a special task team" has been formed to "see what can be done" about W88 modifications.

The congressional request for an independent inquiry was initiated by Rep. John M. Spratt Jr. (D-S.C.), who chairs the House Armed Services Committee's panel on defense nuclear facilities. He was joined by Rep. Les Aspin (D-Wis.), the House Armed Services Committee chairman; Rep. William L. Dickinson (Ala.), the committee's senior Republican; and Rep. Jon L. Kyl (Ariz.), the senior Republican on the committee's defense nuclear facilities panel.

Details of the Trident safety problem described in this article were not obtained from congressional sources.

Watkins said that in early 1989 he told senior aides to ask the directors of the nation's three nuclear weapons laboratories whether they were "satisfied" with the agency's handling of safety issues. "The answer was no," he said, and one lab director used the opportunity to raise strong concerns about a particular weapon that "needed to have some aggressive attention."

Watkins declined to say what the weapon was, but officials at another agency identified it as the SRAM-A, a 1970s-vintage weapon carrying a warhead that uses the same volatile material as the W88 Trident warhead. Roughly 14 feet in length, the weapon can be slung below the wings of B-52, B-1B or FB-111 bombers or carried in the internal bomb bay. "It's basically a fuel tank with wings," one official said.

They said longstanding safety concerns about the weapon are partly based on an intense B-52 engine fire on the runway at Grand Forks, N.D., in September 1980 that injured a crewman and came close to causing an electrical short in the SRAM. The short might have caused the volatile material to explode, dispersing the plutonium in the weapon's core, several officials said.

Acting at Watkins's initiative, DOE officials sought to mention several safety problems involving the SRAM-A warhead in a routine report to Bush last year about the overall safety of the nuclear weapons stockpile. But DoD officials rebelled, causing submission of the report to the White House to be held up for more than three months, according to officials at both agencies.

Watkins said he used the dispute to win DOD's approval for a new weapons safety review committee under DOE's control. He also said the safety matters at issue were explained to Bush by national security adviser Brent Scowcroft with Cheney's concurrence.

"I would have just moved unilaterally [with Bush] had I not been satisfied that the thing was being well aired," Watkins said. Other officials said Watkins and Cheney agreed on the need to control aircraft operations involving the SRAM-A tightly while further analysis is being done.

DoD spokesman Williams said "the Joint Staff did approve modifications to procedures involving the SRAM-A," but declined to say what they were.

Notes

1. Sidney Drell, "Designing and Building Nuclear Weapons to Meet High Safety Standards," chapter 1, this volume.

2. 1988 DOE Nuclear Weapons Safety Review Group, Gordon Moe, chairman.

3. Sidney D. Drell, John S. Foster Jr., and Charles H. Townes, "The report of the nuclear weapons safety panel: hearing before the Committee on Armed Services," House of Representatives, 101st Cong., 2nd sess., December 18, 1990, vol. 5 (Washington, D.C.: U.S. Government Printing Office, 1990).

The Interplay Between Civilian and Military Nuclear Risk Assessment, and Sobering Lessons from Fukushima and the Space Shuttle

3

CHRISTOPHER STUBBS

Introduction

Nations strive to strike an informed balance in the nuclear domain among risk, adverse consequences, cost, and benefit. The two most challenging facets of this for the United States are nuclear power generation (both for electricity and for marine propulsion) and the management of the nation's nuclear arsenal. This paper explores the inter-relationships of risk assessment, risk perception, and risk management between the civilian and military sectors, and draws upon lessons that can be elicited from recent events in Fukushima and from the space shuttle program.

Our nuclear challenge is to achieve the optimum balance among cost, risk, consequences, and benefit. But the optimal solution changes over time. This goal is a moving target because of many factors, including the end of the Cold War and other geopolitical changes—in particular the increase in the dangers of nuclear weapons and proliferation accompanied by the demise of the Soviet Union, and with it the fading of fear of a nuclear holocaust. Other factors include varying concerns about energy security and environmental factors; changing public attitudes and priorities; and the evolution of fiscal realities. The relative weights accorded the various factors that bear upon optimal utilization of nuclear technology differ between the civil and military sectors in the United States; furthermore, different nation states have come to widely differing conclusions about where their nuclear optima reside. In the post-Fukushima world we have seen some examples of radical changes in national nuclear policies.

A quantitative framework for risk assessment has been developed and applied to both the civilian and military nuclear domains. One of the main points of this paper is that the space shuttle program and the Fukushima experience provide sobering reminders of the limitations of probabilistic risk appraisals. Trying to assess the risks associated with highly unlikely scenarios, but which have excruciatingly adverse consequences, is an intrinsically tough problem. Since performance-based data are sparse in evaluating highly unlikely scenarios, it is very difficult to develop a quantitative measure of confidence in assessing relative risk levels, much less absolute ones.

It is important to understand both the strengths and limitations of probabilistic risk assessment, given the important role it plays in the management of complex, high-consequence nuclear systems. There is a long history of using probability and statistics to quantify the risk aspects of nuclear power plants and also to understand the safety and surety of nuclear weapons. In the United States, the civilian side (through the Nuclear Regulatory Commission) and the military side (through the Department of Defense and the National Nuclear Security Administration of the Department of Energy) phrase the safety requirements they impose on nuclear systems in the context of acceptable probabilities.

Section 2 presents the evolution of risk assessment for the space shuttle program as an example of some of the pitfalls of quantitative risk assessment. Section 3 explores in more detail the concepts and challenges of probabilistic risk assessment. This is followed in section 4 by a discussion of the Fukushima nuclear accidents and some speculation about the saturation-level media coverage they received. Section 5 closes the paper with some conclusions and some questions that can help inform the interpretation and critical appraisal of probabilistic pronouncements.

Probabilistic Risk Assessment Lesson 1: The Space Shuttle Program

The space shuttle program has now ended, and provides us with a valuable end-to-end lesson in risk assessment, tempered by the actual experience with the program. The National Aeronautics and Space Administration has undertaken an extensive and exhaustive program of risk assessment for the manned space flight program, so this is a reasonable proxy for the challenges we face in the nuclear domain.

The investigation that followed the 1986 Challenger disaster paid much attention to the discrepancy between risk assessments made by engineers as opposed to program managers, assisted by Richard Feynman's memorable demonstration of the thermal effect on O-rings. But we now know much more about the evolution of the risk assessment and ongoing improvements to flight safety during the course of the shuttle program. Table 3.1 shows the remarkable evolution of NASA's understanding of the shuttle flight risks.

Two important lessons emerge from a recent retrospective appraisal[1] of shuttle program risk:

Table 3.1 Evolution of NASA's assessment of the likelihood of loss of vehicle and crew, for shuttle program. Note that the re-assessment at the end of the program indicates that even the engineering 1980s estimates were likely optimistic *by a factor of ten*, and that there was (according to NASA retrospective analysis) only a 6 percent chance of making it through the first twenty-five flights without a catastrophic accident.

1981	First orbital mission. Initial risk assessments: Engineers: 1:100 Management: 1:100,000
1986	Challenger lost, 25th shuttle flight
1988	1:55
1993	1:73
1995	1:131
1998	1:234
2003	1:78
2003	Columbia lost, 113th shuttle flight
2004	1:61
2005	1:67
2006	1:77
2008	1:81
2009	1:85
2010	1:89
2011 re-analysis of 1980s risks	1:10
2011	Final mission, flight number 135.
Actual	2:135=1:67

1. The engineers' risk assessments in the early days of the shuttle program were overly optimistic by a factor of ten! With the benefit of hindsight, NASA now thinks the likelihood of losing a vehicle and crew for the first shuttle launches was about 10 percent per flight. Safety improved rapidly after the first few flights, but the important point here is that the assessments made by engineers at the time were 1 percent. (With two losses over the course of 135 missions, the actual figure was 1 in 67, or 1.5 percent.)

2. The second lesson pertains to the confidence ascribed to these risk estimates. The most recent NASA risk assessment[2] for the shuttle shows a remarkable level of confidence in its estimate of likelihood of loss of vehicle per launch at the end of the program, given (1) above. One wonders whether the current risk estimates have a realistic systematic uncertainty. NASA shuttle program manager John Shannon remarked, "The instructive piece of this is that over thirty years of operations, two accidents, countless engineering tests and all those things—looking back at it, (now) we understand what the real risk was. But there was no way to know at the time."[3]

The U.S. space shuttles were designed long ago and employed obsolete technology in a complex system, operated by an extensive federal agency that was charged with their safe operation in the national interest. There are clear analogies to both the civilian and military nuclear sectors. The unappreciated systematic errors and early underestimates in the extensive risk assessments carried out by NASA stand as a sobering testament to the difficulty of achieving accurate modeling of complex systems.

Quantifying Nuclear Risk through Probabilistic Risk Assessment

We crave some quantitative framework for risk assessment, in both relative and absolute terms, to help structure our thinking about these issues. Statistical methods and vocabulary are now the standard approach, under the framework of "probabilistic risk assessment." This is a formal methodology that attempts to estimate the likelihood of certain situations, and to fold those appraisals into an overall judgment about the safety margins associated with complex technical systems.

The most demanding quantitative risk assessment problems are those that have high consequences for failure, that contain complex systems of systems with a combination of sophisticated hardware and software, and that include humans in short-time-scale critical decisions. Clearly, the civilian and military nuclear systems satisfy these conditions.

We should distinguish at this point between *relative* risk assessments, which can guide investments in risk reduction and resource allocation, and *absolute* risk assessments that can guide broader policy considerations and implementation decisions. Relative and absolute risk assessments are applicable to both military and civilian nuclear issues.

Entire journals are devoted to publishing academic papers that pertain to risk assessment and risk management (e.g., *Journal of Contin-*

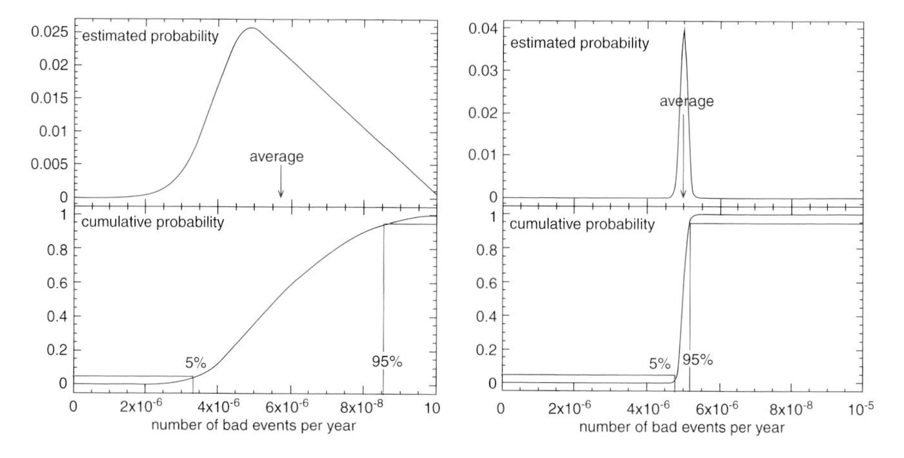

Figure 3.1 The basic concepts of probabilistic risk assessment. The upper panels show two possible outcomes of predictions of rare undesirable events. One could for example imagine the upper graphs represent the estimated time-to-failure of some component or system. Both examples have about the same maximum in their probability distribution, i.e., the peaks in the upper panels. But they have different *average* predicted incidence rates, with the one on the left being about 16 percent higher than the one on the right, as shown by the arrows. But more importantly, note the difference in the *shapes* of the two probability distributions. The prediction on the right is much more concentrated than that shown on the left. The lower panels show the respective "cumulative" probability distributions, from which we can derive confidence bounds. The red lines in the lower panels indicate the 5th and 95th percentile confidence predictions of the event rate. This means that there is a 5 percent chance that the actual rate is less than the 5 percent bound or above the 95 percent bound. For the example on the left, we would assert with 95 percent confidence that the number of events per year is less than 8.5×10^{-6}, that the average of the prediction is 5.8×10^{-6}, and that with 5 percent confidence the number of events is less than 3.4×10^{-6} per year. The example on the right has much more tightly packed confidence levels, which indicates a more secure prediction of the underlying rate. The spread between the 5th and 95th percentile lines is accordingly much smaller. The interpretation of statistical risk predictions must take the statistical spread of the predictions into account, not just the average.

gencies and Crisis Management and *Risk Analysis*). The probabilistic risk assessment methodology has been employed for numerous systems, including air traffic control, space flight (both manned and unmanned), nuclear weapons, nuclear power plants, and health care. Numerous books have explored both the technical and organizational cultural aspects of the nuclear enterprise.

Figure 3.1 illustrates some of the basic quantitative concepts of probabilistic risk assessment.

It is interesting to note that the high-level, unclassified, quantitative probabilistic DoD requirements do not appear to have an associated statistical confidence of the kind shown in Figure 3.1. Is the nation's nuclear complex supposed to assure with 95 percent confidence that such a detonation won't occur, or is this interpreted as an *average* expectation? As discussed below and illustrated in Figure 3.1, our understanding of probability *distributions* is an essential aspect of informed risk management and nuclear decision-making.

The spread in NASA's assessment[4] of crew and vehicle loss likelihood for the final shuttle flights was astonishingly narrow. There is only a factor of two difference between the 5 percent and 95 percent confidence intervals. I find this hard to understand, given NASA's judgment that prior estimates were off by a factor of ten!

In many ways the probabilistic assessment of risk is a very mature discipline. However, if the probabilistic risk assessment methodology were working properly we would not have been taken by surprise by the events at Fukushima, nor by the two space shuttle accidents.

Quantitative Risk Assessment and Requirements, Nuclear Weapons

The nuclear weapons complex sets forth very stringent quantitative requirements for nuclear weapons. The origin and effectiveness of these requirements are explored by Sidney Drell in chapter 1 and Robert Peurifoy in chapter 2. Under stressing conditions but prior to active use, the Defense Department's requirement stipulates that the chance of achieving nuclear detonation be less than one part per million. It happens that this coincides with the NRC's expectations for about 10^{-6} nuclear accidents per year of reactor operation.

Quantitative Risk Assessments in Civilian Nuclear Systems: The Nuclear Regulatory Commission and Its Predecessors

The civilian nuclear sector has a long history of struggling with probabilistic risk assessment and its appropriate interpretation. The 1975 Rasmussen report[5] was an important early document that led to considerable push-back from members of the American Physical Society. The basic framework of fault tree modeling and the defense-in-depth concept (discussed in more detail below) were both established in that era.

The quantitative requirement from the NRC puts risk from reactor accidents as a relative marginal risk above the "risk background," as noted in a 1986 policy statement:[6] "The risk to an average individual in the vicinity of a nuclear power plant of prompt fatalities that might result from reactor accidents should not exceed one-tenth of one percent (0.1 percent) of the sum of prompt fatality risks resulting from other accidents to which members of the U.S. population are generally exposed."

The defense-in-depth concept for civilian nuclear power plants relies on layered protective systems, such as reactor shutdown systems and containment buildings. These systems are engineered to be independent, in the sense that their failure probabilities are also independent of each other. This allows two independent systems that each have a part-per-thousand failure likelihood to provide a joint failure probability that is the product of these, or a part per million. But this assumption of independence must be justified, especially under the most severe external challenge to the system's integrity.

Why Probabilistic Nuclear Risk Modeling Is Hard

Although we really don't have an alternative to probabilistic risk assessment, there are major challenges to making accurate estimates for rare events. Some of these are listed below.

Challenge 1 Identifying and Estimating Correlations

The practice of multiplying probabilities together to obtain a combined likelihood of failure only works if the events being modeled are truly independent. Any correlated failure mode undermines any estimate that ignores these connections. It's difficult to identify and estimate subtle correlations.

Challenge 2 Non-Gaussian Statistics

There is often an implicit assumption that the events being modeled obey Gaussian statistics. Complex and non-linear systems in the real world usually exhibit non-Gaussian behavior. Rare events in the unexplored non-Gaussian tails are difficult to estimate, especially when we have a very limited experience base in this domain.

Challenge 3 Accurately Modeling Human Factors

It is notoriously difficult to accurately incorporate the interplay between people and complex technical systems, especially under adverse conditions. Experts who have investigated the sequence of events at Fukushima acknowledge that a different set of decisions by the operators could have produced a different outcome. Per Peterson, chairman of nuclear engineering at the University of California-Berkeley, was quoted[7] by National Public Radio: "It's quite likely that if the injection of seawater had been initiated earlier, the damage of fuel could have been limited greatly or even prevented. . . . So I think there are possible pathways by which the severity of the accident could have been substantially less."

Human factors play a critical role in the management of nuclear systems. The training, demeanor, and experience of the individuals entrusted with these systems are critical determining factors in their ability to cope with unexpected situations.

A literally textbook example of a culture of elite excellence is the U.S. nuclear navy. This group is responsible for both the operation of marine nuclear propulsion plants and the operation of the navy's nuclear weapons at sea and ashore. Two U.S. nuclear submarines have been lost at sea (the USS *Thresher* in 1963 and the USS *Scorpion* in 1968) but both are thought to be non-nuclear accidents. By contrast, there have been twenty reported reactor-related incidents on Soviet or Russian subs and ships from 1960 to 1993.[8] Assuming comparable fleet sizes, this is a significant difference.

The culture of the U.S. nuclear navy is attributed to Admiral Hyman Rickover and his strong hands-on management of the entire program. Rickover was a participant in the Manhattan Project,[9] so there is a direct link between the early nuclear weapons program culture and that of the navy's nuclear propulsion program. The ongoing transfer of trained and experienced personnel from the nuclear navy into civilian reactor operations is a lateral transfer of skills and outlook from the military sector into the civil nuclear power arena.

Challenge 4 Narrow Base of Experience

We are working from a limited experience base, and we make extrapolations. Probabilistic risk assessment is an attempt to gain understanding in a regime where we don't typically have direct experience. This necessarily means making an informed extrapolation, using some kind of statistical model. We should recognize that these extrapolations are fraught with uncertainties that are difficult to accurately estimate.

Challenge 5 Extrapolations Are Based on Assumptions

Probabilistic outputs depend on input assumptions. Consider this statement[10] from the NRC's post-Fukushima report: "The accident in

Japan was caused by a natural event (i.e., tsunami) which was far more severe than the design basis for the Fukushima Dai-ichi Nuclear Power Plant."

Fault tree analysis is a combination of engineering appraisals of component and sub-system failure rates in conjunction with an assessment of the likelihood of various stressing events. The NRC approach is to assume that a representative stressing external forcing function can provide insight into other analogous situations. But the *assumed rate* of these external driving events is typically an extrapolation from existing data sets. It's not my intention to criticize the assumptions that are commonly made, but rather to point out how difficult it is to make shrewd choices.

A topical example that applies to both military nuclear systems and civilian power plants is the terrorist threat. Are we more likely to see an armed assault from a technologically unsophisticated mob on the security perimeter of a facility, or an attack from a knowledgeable insider, or a cyber attack on our control systems? And apart from the *relative* likelihood of these scenarios, what is the absolute probability, expressed in units of attacks per facility per decade? Finally, what about situations that we weren't clever enough to incorporate into our probabilistic risk models? We simply don't know the right answers to these questions. The assumptions we make will greatly influence the risk quantification results produced by statistical models.

Probabilistic Risk Assessment Lesson 2: Fukushima

The unfolding of the nuclear saga at Fukushima following the earthquake and tsunami is an important learning opportunity for increasing our understanding of risk management in *all* sectors of the nuclear arena.

The unfolding events in and around the Fukushima reactors were not anticipated by the probabilistic analyses undertaken prior to the earthquake and tsunami. The power utility, the Japanese government, and the public were taken by surprise. The combination of a natural disaster and consequent technical failures, compounded by human factors, led to a situation that exceeded the range of scenarios considered as plausible insults to the plant.

This is a sobering lesson in the limitations of probabilistic risk management, for all the reasons outlined above.

Perhaps the biggest impact of the Fukushima experience, however, is the erosion of the public's confidence in the ability of governments and power utilities to properly manage the risks associated with nuclear technology. To date this has primarily affected the civilian nuclear power sector, but there is a real prospect of the same apprehensions and re-evaluations extending into the military sector.

The Management of Public Risk Perception

The choices that people make are driven in part by their perception of risk, which is often at odds with the data. The number of automobile deaths per year in the United States is 30,000, but many people are more apprehensive about air travel, which claims only about 1,000 lives per year, *worldwide*.

Risk perception is a complicated topic, and media coverage plays a role. A single traffic accident with ten fatalities typically generates more news coverage than ten single-fatality incidents. Perhaps this is because our interpretation shifts from people as active participants (who made individual decisions to drive) to innocent victims.

The media coverage of the Fukushima nuclear event, in comparison to the devastation due to the earthquake and tsunami, is an interesting contrast. The natural disaster led to devastating loss of life, with an economic impact in the billions of dollars. Yet the media coverage focused on the ongoing nuclear risks from the power reactors. One might speculate that there is a perception that a natural disaster is beyond our control, while the risk to lives and livelihoods by a potential man-made nuclear disaster could have been avoided.

The Public Perception of Risk Management

The flip side of this issue is the public's perception of how well existing organizational structures are managing the risks of nuclear technology and striking the right balance among cost, risk, benefit, and potential consequences.

Germany has decided to shut down all nuclear power plants by 2020. This is despite the fact that even now, "German households pay twice as much for power than in France, where 80 percent of energy is generated by nuclear plants."[11]

We therefore already have direct evidence of the Fukushima events (or perhaps the *perception* of the Fukushima events) driving a reappraisal of the nuclear optimization problem in the civilian power sector. Europe's largest economy has elected to phase out nuclear power plants, apparently even without conducting an in-depth economic, technical, and political analysis. We don't yet know how far-reaching this reassessment will be, and whether it will eventually extend from the civilian into the military nuclear domain.

Conclusions

1. Probabilistic risk assessment is employed in both the civilian and military nuclear sectors. Recent experience (from NASA's space shuttle program and from Fukushima) shows that systematic errors can afflict these risk estimates.
2. It is important to critically re-assess nuclear probabilistic risk analysis for both civilian and military nuclear issues. Some questions that journalists might want to pursue include:
   a. What assumptions were made about external insults to the system?
   b. What correlations between parameters were assumed?

c. What is the probability distribution of the analysis? What are the 5 percent and 95 percent confidence bounds?
d. Were the underlying distributions assumed to be Gaussian? If so, on what basis?
e. How were human factors modeled, and with what uncertainty?
f. What extrapolations were made from validated data?
g. What is the worst-case scenario evaluated, and how was this chosen?
h. What systematic errors are included in the analysis?
i. What external validation and verification methods were applied?

3. Changes in circumstances (e.g., end of the Cold War, terrorist threats, Fukushima) should prompt a re-evaluation of the optimization choices for nuclear technology in both the civilian and military domains.
4. The public's perception of risk and its associated confidence in the governmental and private sectors' ability to appropriately manage risk on its behalf have been adversely affected by the Fukushima events.

Acknowledgments

I am grateful to the Hoover Institution for the Annenberg Fellowship that allowed me to attend and contribute to this conference, and in particular for the remarkable intellectual climate provided by George Shultz and Sid Drell. I would also like to extend my thanks to Barbara Egbert, whose expert and professional assistance greatly improved this manuscript.

Notes

1. Roger L. Boyer, "Space Shuttle Probabilistic Risk Assessment, Iteration 3.2," presented at Trilateral Safety and Mission Assurance Conference, October

26–28, 2010, Cleveland, Ohio, http://ntrs.nasa.gov/archive/nasa/casi.ntrs.nasa.gov/20100036684_2010040595.pdf.

2. Ibid.

3. Todd Halvorson, "Analysis: NASA underestimated shuttle dangers," *Florida Today,* February 13, 2011, http://www.usatoday.com/tech/science/space/2011-02-13-nasa-underestimated-risk_N.htm.

4. Boyer, "Space Shuttle Probabilistic Risk Assessment, Iteration 3.2."

5. U.S. Nuclear Regulatory Commission, "WASH-1400. Reactor Safety Study. An Assessment of Accident Risks in U.S. Commercial Nuclear Power Plants." (Study was directed by Professor Norman C. Rasmussen, Massachusetts Institute of Technology.) http://teams.epri.com/PRA/Big%20List%20of%20PRA%20Documents/WASH-1400/02-Main%20Report.pdf.

6. Nuclear Regulatory Commission, "Safety Goals for the Operation of Nuclear Power Plants," policy statement, *Federal Register,* vol. 51, 1986: p. 30028, http://www.nrc.gov/reading-rm/doc-collections/commission/policy/51fr30028.pdf.

7. Richard Harris, "What Went Wrong In Fukushima: The Human Factor," *NPR online,* July 5, 2011, http://www.npr.org/2011/07/05/137611026/what-went-wrong-in-fukushima-the-human-factor.

8. P. L. Olgaard, "Nuclear Ship Accidents, Description and Analysis," Technical University of Denmark white paper (1993).

9. Paul E. Bierly III and J. C. Spender, "Culture and high reliability organizations: the case of the nuclear submarine," *Journal of Management,* vol. 21, no. 4 (1995).

10. Charles Miller, Amy Cubbage, Daniel Dorman, Jack Grobe, Gary Holahan, and Nathan Sanfilippo, "Recommendations for Enhancing Reactor Safety in the 21st Century: The Near-term Task Force Review of Insights from the Fukushima Dai-ichi Accident," Nuclear Regulatory Commission, July 12, 2011, http://pbadupws.nrc.gov/docs/ML1118/ML111861807.pdf.

11. Tom Bawden, "German nuclear shutdown forces E.ON to cut 11,000 staff," *The Guardian,* August 10, 2011, http://www.guardian.co.uk/business/2011/aug/10/german-nuclear-shutdown-forces-eon-to-axe-11000-jobs.

Long-Range Effects of Nuclear Disasters

4

RAYMOND JEANLOZ

In addition to local devastation at the site of conflict, a limited (regional) nuclear war would have long-range health effects that could greatly exceed those of the worst nuclear power-plant accidents to date, with radioactive fallout potentially causing tens of thousands of additional deaths at distances up to thousands of kilometers from the nuclear explosions.

Introduction

The fear of a major nuclear exchange that characterized the Cold War has now receded, leaving the perception that the military use of nuclear weapons is most likely—if at all—in a regional war, with more limited total yield than was anticipated from an exchange between global superpowers. It is also conceivable that a terrorist organization could detonate one or a few nuclear devices[1] (e.g., Garwin, 2010; Buddemeier, 2010). Yet another example of "limited" nuclear conflict might include a nation having a small nuclear arsenal attacking a major nuclear power (or one of its allies).

In any of these cases there would be local devastation, with the potential for unprecedented casualties if one or more nuclear explosions took place in the high population densities typical of modern megacities (NRC, 2005a; Toon, et al., 2007). Much less certain, however, are the consequences at greater distances, around the world, of such limited nuclear war.

The purpose of this article is to address the question: "How limited are the consequences of 'limited' nuclear conflict?" One recently examined aspect of this question involves the potential impact of a regional nuclear exchange on Earth's atmosphere, hence global climate (Robock, et al., 2007). Here the focus is instead on radioactive fallout, taking into account the 1986 Chernobyl and 2011 Fukushima accidents and public reaction to these civilian nuclear-power plant disasters.[2] In short, how does the potential danger due to nuclear weapons compare with that due to nuclear power for those at a distance from the crisis?

Radioactive Fallout

The problem is complicated by at least three factors (see also Stubbs, 2011). First, the effects of nuclear detonations are technically difficult to predict because the underlying physical, chemical, and biological processes are complex. Second, the results are uncertain because of (fortunately) limited experience, and also because of fundamental unpredictability in many of the processes involved (e.g., the effectiveness of long-range particulate and gas transport by wind currents at different heights in the atmosphere). In addition to these uncertainties due to technical considerations, there are inevitably large uncertainties regarding plausible scenarios for regional nuclear war. And, third, there is considerable public skepticism about the reliability of statements made by technical experts, especially if made on behalf of government or industry. That is, technical considerations may have only limited influence on public reaction and, therefore, on the societal consequences of a nuclear disaster.

Black Rain

Skepticism is understandable, given the historical record of inaccurate predictions regarding the consequences of nuclear explosions. For example, the expectations that both Hiroshima and Nagasaki explosions would not leave fallout on the ground, because both bombs were detonated above the "fallout-free" height of burst,[3] were overturned by the experience of "black rain" bringing radioactive material down to ground level.[4] Moreover, the amounts were significant, with accumulated doses[5] more than 100 times higher than the residual radiation[6] present much closer to each of the atomic-bombing hypocenters—e.g., ten to 400 mGy (milligray) at three kilometers from the hypocenter due to black rain, versus 0.2–0.6 mGy residual after one day at one kilometer distance from the hypocenter (Imanaka, 2006; Cullings, et al., 2006) (Figure 4.1).[7]

Radioactive Debris from Nuclear Explosions and Nuclear Power Plants

Ironically, these numbers illustrate why nuclear-weapon specialists consider fallout to be relatively inconsequential in comparison with the immediate effects of the explosion, assuming detonation above the fallout-free height of burst. Residual radiation doses (after one day) of fifty-five to 190 mGy at the hypocenters of the Hiroshima and Nagasaki explosions, and less than one mGy at one kilometer distance, are far smaller than the 350 mGy equivalent lifetime dose that was used as a criterion for relocating populations from the Chernobyl area or the total equivalent dose of 250 mGy to which Fukushima workers were limited after that disaster.

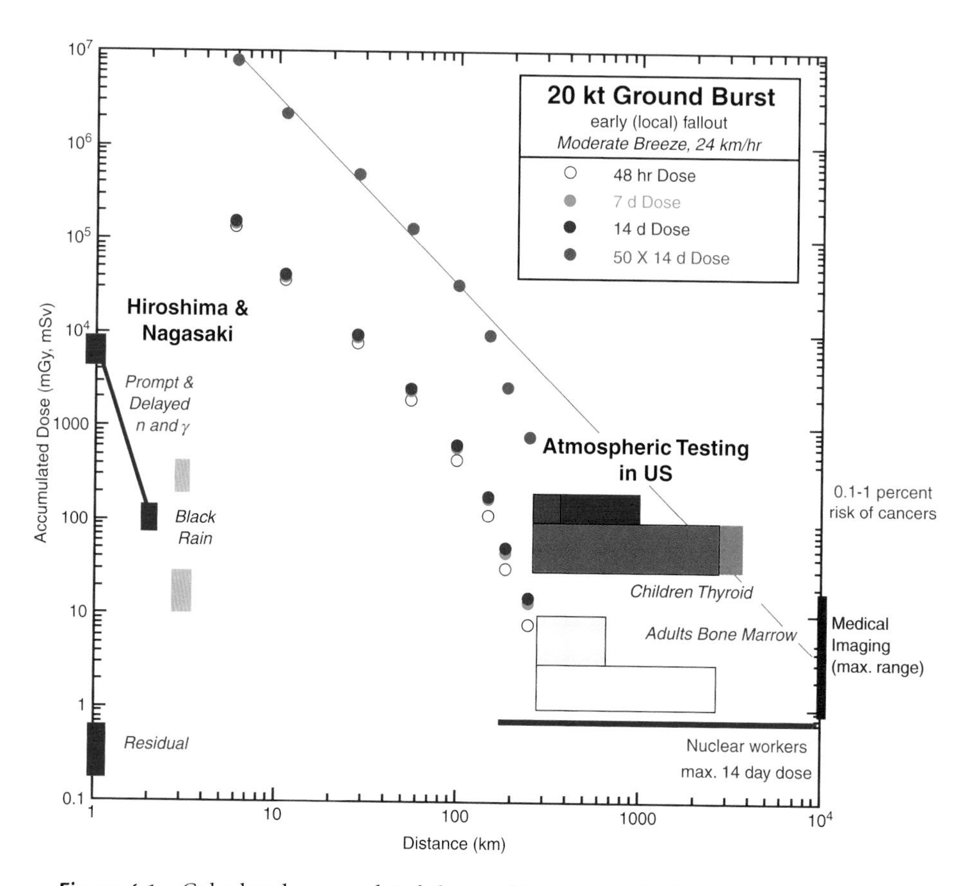

Figure 4.1 Calculated accumulated doses of ionizing radiation due to early (local) radioactive fallout from one 20-kt ground-level nuclear explosion (open, blue, and green points for 48 hour, 7 day, and 14 day doses) and fifty 20-kt nuclear ground bursts (orange points: 14-day dose for 1 Mt total fission yield) are compared with doses estimated for the Hiroshima and Nagasaki bombings (red boxes for prompt and delayed doses due to neutrons and γ-rays at distances of 1–2 km from hypocenter; dark red box for residual radiation at 1 km from hypocenter; grey boxes for "black rain" at 3 km from hypocenter) and for estimated doses to the U.S. population resulting from atmospheric testing at NTS (yellow and orange-red boxes for burdens to adult bone marrow and children's thyroids: see text and note 16). The maximum burden out to ranges of 2,000–3,500 km due to delayed (long-range) radioactive fallout from NTS is roughly consistent with an extrapolation of early fallout out to 60–100 km distance varying as $1/\text{distance}^2$ (thin orange line); note that the east and northeast coast of the United States is at 3,000–4,000 km distance from NTS, so no information is available on accumulated doses beyond this range. For reference, the following dose levels are shown on the right: doses associated with 0.1–1 percent increased risk of cancer; maximum doses expected from certain medical imaging procedures (e.g., abdominal CT scan); and 14-day dose limit for nuclear workers.

The point is that nuclear power plants contain large amounts of nuclear fuel-producing fission products for long periods of time, including radioactive materials with decay times of weeks to years (Table 4A.1 in appendix), so these can become important sources of dispersible radioactive debris (e.g., NRC, 2005a, p. 92). In comparison, a nuclear weapon is relatively small and inefficient in producing fission products, and many of those that are produced decay rapidly.[8]

Therefore, for an explosion above the fallout-free height of burst and no precipitation, a nuclear weapon produces little fallout. The most extreme doses are experienced as *prompt* radiation (i.e., radiation due to particles and rays emitted directly from, and experienced immediately after, the explosion) rather than fallout, with greatest intensities near the hypocenter where blast and thermal effects predominate. This particular fact re-emphasizes the distinction between nuclear power and nuclear weapons: the immediate effects of blast, heat, and radiation from a nuclear explosion are enormously more damaging than those of a power-plant accident.

Nuclear Conflict

Whereas nuclear-explosion testing was largely conducted in a manner that minimizes production of fallout, the same may not hold for nuclear conflict. For example, terrorist detonation of a nuclear device would almost certainly be near ground level, thereby greatly enhancing the production of radioactive fallout as compared with the high airbursts discussed above (see note 3). From a military perspective, the ability to destroy hardened facilities, especially if buried (e.g., bunkers or control centers), is significantly enhanced by detonating a nuclear weapon at or below ground level (NRC, 2005a). More speculatively, it would be prudent to anticipate ground bursts in a regional nuclear war, if for no other reason than error during conflict, let alone intentional targeting.

Applying well-established methods of estimating the initial (local) fallout from nuclear weapons, one finds a dose rate of 3,000 mGy/h after one hour (h) at a distance of about twenty-five kilometers from the site of a twenty-kiloton (kt) ground burst, comparable to the yield of the

Nagasaki explosion (Glasstone and Dolan, 1977; NRC, 2005a).[9] In comparison, dose rates of only 0.3 mGy/h were determined on the edge of the twenty-kilometer Fukushima exclusion zone and peak dose rates measured on the edge of the plant remained below about ten mGy/h. For perspective, 0.1 mGy/h is twenty times the maximum rate measured in Finland after the Chernobyl accident.[10]

These considerations apply to a single explosion and do not address the consequences of an actual conflict. It is not implausible that regional conflicts might employ weapons having fission yields in the twenty-kiloton range. However, the number of such weapons that would be used is extremely uncertain. For purposes of discussion, one megaton (Mt) total fission yield is considered here, amounting to just over two dozen twenty-kiloton weapons being detonated by each side in an exchange (cf. Toon, et al., 2007).

The result of such a calculation gives fourteen-day accumulated doses exceeding one, ten, and one hundred mGy out to distances of about 650, 480, and 350 kilometers for the early fallout from fifty twenty-kiloton detonations (Figure 4.1).[11] For comparison, a one hundred-mGy dose-equivalent is associated with a risk of cancer of 0.1 percent, and the maximum average nuclear worker's dose limit is under one mGy for fourteen days.[12]

Uncertainties

There are many uncertainties in this analysis, and its only purpose is to provide an idea of the possible magnitudes involved. Terrain effects can reduce total doses by factors of 0.3–0.5, for instance, and precipitation as well as local winds cause heterogeneous distributions of fallout (e.g., radiation "hot spots" present within large regions of reduced average dose); winds could also have higher velocities, dispersing the radiation farther than assumed here (Glasstone and Dolan, 1977). Crucially, it is not obvious that one megaton worth of ground bursts would characterize an actual regional conflict.

In a sense, however, this approach gives an underestimate, in that delayed (long-range) fallout associated with longer-lived radionuclides is considered even more difficult to predict, so is not taken into account (Glasstone and Dolan, 1977). Therefore, instead of modeling one can turn to experience gained from atmospheric nuclear-explosion testing in order to gauge the long-range spread of fallout. In fact, the one megaton calculation is relevant for comparative purposes because the cumulative fission yield of U.S. atmospheric nuclear explosions at the Nevada Test Site (NTS)[13] amounted to approximately one megaton between 1951 and 1962.

Atmospheric Testing

There has been considerable analysis of the spatial distribution of doses resulting from NTS tests, including the radioactivity that was lofted to great distances. In particular, a 1997 National Cancer Institute (NCI) study established representative dose levels and epidemiology, with findings that have been largely validated by subsequent reviews (e.g., NRC, 1999; Beck and Bennet, 2005; NRC 2005b; DHHS, 2005) and by continued technical study (Bennet, 2002; Bouville, et al., 2002; Gilbert, et al., 2002; Simon and Bouville, 2002; Simon, et al., 2004; Beck, et al., 2006; Gilbert, et al., 2010). The health consequences have been a focus of special attention because of RECA[14] compensation to more than 15,000 "downwinders" and others affected by fallout from NTS, to date (NRC, 2005b).[15]

The NCI results include estimates of accumulated doses ranging from one to three mGy out to distances of 2,400–3,700 kilometers and three to ten mGy to 650–850 kilometers for total internal and external burden to the red bone marrow of adults; and of thirty to one hundred mGy out to 2,600–3,700 kilometers and more than one hundred mGy to 350–1,000 kilometers for total internal and external doses to the thyroids of milk-drinking children (Figures 4.1–4.4 of NRC, 2005b).[16]

The studies conclude that external exposure to ionizing radiation due to nuclear explosions at NTS amounted to about 79,000 person-Gy (26,000–237,000 person-Gy)[17] (DHHS, 2005: Table 3.14). Correcting for air and balloon bursts,[18] and using a lifetime attributable risk mortality coefficient of 0.06/Gy,[19] implies 6,300 (2,100–19,000) extra fatalities due to external exposure from one megaton of ground-burst nuclear explosions, with twice as many (approximately 12,600) cancer casualties as fatalities. In addition, 49,000 (11,300–212,000) lifetime excess thyroid cancers are expected due to internal exposure from the NTS fallout (NRC, 1999: Appendix B); with a thirty-year mortality rate of 10 percent,[20] this implies 4,900 (1,100–21,200) extra fatalities due to thyroid cancers.

In total, therefore, experience at NTS suggests that approximately 11,000 additional fatalities can be associated with one megaton (total fission yield) of nuclear explosions at ground level for a population of about 150 million extending approximately 2,000 to 4,000 kilometers downwind.[21] These values are comparable to the effects on the U.S. population expected from global atmospheric nuclear-explosion testing (primarily U.S. and Soviet): 22,000 radiation-related solid cancers, with half being fatal; and radiation-related leukemia that might eventually cause a total of about 1,800 deaths (NRC, 1999; DHHS, 2005; Gilbert, et al., 2002; cf. Beck and Krey, 1983).[22]

To summarize, findings of radioactive fallout extending thousands of kilometers from NTS and of thousands of additional deaths potentially being associated with nuclear-explosion testing are striking. These results suggest that the classical modeling of radioactive fallout from nuclear detonations—with a focus on initial, local fallout—may be more of an underestimate than previously recognized.

Societal Impact

The figures for fallout-attributable fatalities must be kept in perspective, in that they amount to less than about 0.1 percent of deaths expected for the same populations due to baseline cancers or leukemia. It is small comfort to those fearing they are among the many thousands of *additional* casualties due to radioactive fallout, however, to be told that their cohort is in any case expected to experience 60 million deaths from cancer and 1.5 million deaths from leukemia.[23] Similarly, it does no good to point out that certain medical imaging procedures can impose doses of as much as one to twenty mGy, comparable to the adult bone-marrow burden from NTS, because such procedures are undertaken as a matter of choice.

Radioactive fallout has systematically greater health effects on women and children relative to men and adults, respectively (BEIR-VII: NRC, 2006). In addition, the effects on animals and plants are not considered in the present discussion, even though there can be significant impact on food supply (e.g., contamination of crops over wide areas). Another detail that is overlooked here is the likely impact on air travel due to the long-term presence of radioactive particles and gases in the atmosphere.

These points get to the most difficult aspect of the question at hand, namely the public reaction to radioactive fallout. There is little question that what is tolerated as a consequence of personal choice may be intolerable when imposed involuntarily. In combination with the public's concerns about radioactivity, it is then understandable that conservative limits are necessarily imposed on potential exposures to ionizing radiation from power plants: ten to fifty mGy (ten to fifty mSv) per year for nuclear workers and one mGy per year (one mSv/y) above background for the population at large.[24] Reaction to power-plant accidents such as Chernobyl and Fukushima reinforce the distinction between public intolerance of imposed risk and acceptance of comparable risks due to personal choice.

The matter is even starker in considering regional nuclear conflict, because now the fallout—and consequent risk—would be the result of deliberate human action, rather than of unanticipated accident.[25] Transport of radioactive fallout to 2,000 to 4,000 kilometers means that nuclear explosions in—for instance—Israel, the India/Pakistan border region, or the Korean Peninsula could inevitably have health effects extending as far as Pakistan (and including Iran), China (and southeast Asia), and Japan, respectively. Downwind populations could amount to roughly 340 million and more than 900 million within 4,000 kilometers for the first two examples, and 130 million within 2,000 kilometers for the third example. Scaling from the extra fatalities estimated for a one-megaton total fission yield of ground bursts, based on NTS experience with nuclear-explosion testing, leads to an anticipation of approximately 25,000, 62,000, and 9,000 additional deaths due to fallout for these three examples.[26] The fallout encircling the globe would be measurable in near-real time, at least as readily as was the case after Fukushima.

Conclusions

Considerable information has been obtained in the past twenty to thirty years regarding the radioactive fallout which emanated to long distances from nuclear explosions, as well as from accidents occurring at nuclear power plants. Far-field health effects are generally small when considered as a percentage of the affected population, but can readily amount to many thousands (even tens of thousands) in absolute numbers. Public reaction to dispersal of radioactivity has been intense, even when objective evidence of health effects is difficult to establish. It is therefore implausible to expect that the threat of an impending regional nuclear conflict would be considered anything short of a world-wide health emergency, likely outstripping nations' reactions to the Chernobyl and Fukushima power-plant accidents.

Appendix: Radioactivity in Nuclear Explosions and Nuclear Power Plants

That it makes sense to compare fallout from nuclear explosions and nuclear power plants is suggested by the abundances of radioactive decay products present in these two sources (Table 4A.1).[27]

In round numbers, the amount of radioactivity contained in the core of a 0.5 GWt reactor is comparable to that produced by a one- to two-megaton explosion: about 10,000 PBq are present 24 hours after shutdown, following two years of operation (AEC, 1957, p. 7; Eisenbud and Gesell, 1997, pp. 233–236, 250; note 10).[28] Of course, it is implausible that more than a fraction of a reactor core's fission products would be released into the environment due to an accident, whereas all of a nuclear weapon's fission products are discharged in an explosion. Still, it is evident that the release from Chernobyl was a megaton-scale event, whereas Fukushima was smaller.

Table 4A.1 Radioactive Fission Products in Nuclear Explosions and Nuclear Reactors

Radionuclide	Half Life	1 Mt Explosion (PBq)[a]	1 GWt Reactor (PBq)[b]	Chernobyl (PBq)[c]	Fukushima (PBq)[d]
^{89}Sr	51 d	740	888	115	
^{90}Sr	29 y	4	67	(8) 10	
^{95}Zr	64 d	925		84	
^{103}Ru	39 d	685		>168	
^{106}Ru	1 y	11		>73	
^{131}I	8 d	4,625	1,036	(270) 1,760	142
^{137}Cs	30 y	6	89	(37) 85	10–36
^{141}Ce	33 d	1,443		84	
^{144}Ce	284 d	137	1,295	50	
Total		**8,575**		**11,000**	

[a]Table 9-2 in Eisenbud and Gesell (1997); slightly different values are given in the original (Klement, 1959, 1965) as well as in Table 3 of Anspaugh (1996), such that the total radioactivity per 1 Mt fission yield is expected to lie between about 6,400 and 9,700 PBq (see note 27).

[b]Partial listing from Table 8-1 in Eisenbud and Gesell (1997) of the fission products present in a 1 GWt reactor core 24 hours after shutdown following two years of operation (sf. Stohl, et al., 2011: Table 2).

[c]Fission products released in the accident of April, 1986: right column is a *partial* listing from UNSCEAR (2008: Annex D, Table A1), with the total from Eisenbud and Gesell (1997, p. 415). Values in parentheses (left column) are estimates made in 1986 that are now considered unreliable (from Table 12-6 of Eisenbud and Gesell, 1997). Chernobyl Unit 4 was a 3.2 GWt reactor that was destroyed in the accident.

[d] Fission products released in the accident of March, 2011: partial listing from Morino, et al. (2011) and Stohl, et al. (2011).

Nuclear explosions spread radioactive materials from several sources.

1. The initial nuclear explosive materials that are radioactive, such as uranium-235 (^{235}U) or plutonium-239 (^{239}Pu): these are in quantities of order tens of kilograms,[29] but the decay rates are sufficiently slow that the overall activities are relatively limited.[30]
2. The radioactive products of fission: these can be abundant enough and have fast enough decay rates as to significantly affect health (not all products of radioactive decay are radioactive, however).

3. Bomb material that absorbed neutrons generated during the nuclear explosion (neutron activation: see note 3), a kind of self-induced radioactivity.
4. Material in the environment (e.g., rock, soil, air) that absorbed neutrons produced by the nuclear explosion.

These radioactive materials can decay into other radionuclides (some of which may also be radioactive), be absorbed onto a wide variety of material surfaces, and get moved around through air, water, or ground environments. Such radioactive debris (typically, small particles and gas) reaching ground level is considered fallout that can affect human health due to the release of ionizing radiation.

Nuclear power plants similarly contain radioactive fuel, radioactive fission products, and a limited amount of self-induced radioactivity. These are comparable to the first three sources listed above for radioactive materials associated with nuclear explosions (there is no neutron activation of the environment outside the reactor core), and are present in the core of the reactor as well as in spent fuel rods that may be stored outside the reactor (e.g., in cooling pools: see Stohl, et al., 2011, regarding Fukushima). The tons of un-irradiated fuel typically present in a nuclear power plant are often considered a negligible health risk—unless a large amount were widely dispersed—due to the fuel's slow decay rates. In contrast, many of the fission products that have relatively fast decay rates can pose significant health risks if dispersed into the environment in sufficient abundances.

Finally, because the abundances of individual fission products are different for nuclear reactors and explosions, only approximate comparisons can be made between these two potential sources of fallout.

Acknowledgments

I thank M. L. Adams, J. F. Ahearne, T. W. Boyer, S. D. Drell, R. L. Garwin, Z. Geballe, S. L. Simon, and C. Stubbs for helpful discussions.

References

L. R. Anspaugh, 1996, Technical basis for dose reconstruction, *UCRL Report JC-123714*, Lawrence Livermore National Laboratory, Livermore, CA, 38 pp. http://www.osti.gov/bridge/servlets/purl/266654-oQ8Iqm/webviewable/266654.pdf.

L. R. Anspaugh, R. J. Catlin, and M. Goldman, 1988, The global impact of the Chernobyl reactor accident, *Science*, 242, 1513–1519.

Atomic Energy Commission (AEC), 1957, Theoretical possibilities and consequences of major accidents in large nuclear power plants, *Report WASH-740*, U.S. Atomic Energy Commission, Washington, D.C., 105 pp.

H. L. Beck and B. G. Bennett, 2002, Historical overview of atmospheric nuclear weapons testing and estimates of fallout in the continental United States, *Health Physics*, 82, 591–608.

H. L. Beck and P. W. Krey, 1983, Exposures in Utah from Nevada nuclear tests, *Science*, 220, 18–24.

H. L. Beck, L. R. Anspaugh, A. Bouville, and S. L. Simon, 2006, Review of methods of dose estimation for epidemiological studies of the radiological impact of Nevada Test Site and global fallout, *Radiation Research*, 166, 209–218.

H. L. Beck, A. Bouville, B. E. Moroz and S. L. Simon, 2010, Fallout deposition in the Marshall Islands from Bikini and Enewetak nuclear weapons tests, *Health Physics*, 99, 124–142.

B. Bennett, 2002, Worldwide dispersion and deposition of radionuclides produced in atmospheric tests, *Health Physics*, 82, 644–655.

A. Bouville, S. L. Simon, C. W. Miller, H. L. Beck, L. R. Anspaugh, and B. G. Bennett, 2002, Estimates of doses from global fallout, *Health Physics*, 82, 690–705.

A. Bouville, H. L. Beck, and S. L. Simon, 2010, Doses from external irradiation to Marshall Islanders from Bikini and Enewetak nuclear weapons tests, *Health Physics*, 99, 143–156.

B. Buddemeier, 2010, Reducing consequences of a nuclear detonation, *The Bridge*, 40, 28–38.

E. Cardis, et al., 2006, Cancer consequences of the Chernobyl accident: 20 years on, *Journal of Radiologic Protection*, 26, 127–140.

H. M. Clark, 1954, The occurrence of an unusually high-level radioactive rainout in the area of Troy, NY, *Science*, 119, 619–622.

H. M. Cullings, S. Fujita, S. Funamoto, E. J. Grant, G. D. Kerr, and D. L. Preston, 2006, Dose estimation for atomic bomb survivor studies: Its evolution and present status, *Radiation Research*, 166, 219–254.

Department of Health and Human Services (DHHS), 2005, *Report on the Feasibility of a Study of the Health Consequences to the American Population from Nuclear Weapons Tests Conducted by the United States and Other Nations*, Centers for Disease Control and Prevention, Atlanta, GA, http://www.cdc.gov/nceh/radiation/fallout/default.htm.

F. de Vathaire, V. Drozdovitch, P. Brindel, F. Rachedi, J.-L. Boissin, J. Sebbag, L. Shan, F. Bost-Bezeaud, P. Petitdidier, J. Paoaafaite, J. Teuri, J. Iitis, A. Bouville, E. Cardis, C. Hill, and F. Doyon, 2010, Thyroid cancer following nuclear tests in French Polynesia, *British Journal of Cancer*, 103, 1115–1121.

M. Eisenbud and T. Gesell, 1997, *Environmental Radioactivity from Natural, Industrial and Military Sources*, 4th Ed., Academic Press, Elsevier, New York, 639 pp.

R. L. Garwin, 2010, A nuclear explosion in a city or an attack on a nuclear reactor, *The Bridge*, 40, 20–27.

R. L. Garwin and G. Charpak, 2001, *Megawatts and Megatons*, Knopf, New York, 412 pp.

E. S. Gilbert, C. E. Land, and S. L. Simon, 2002, Health effects from fallout, *Health Physics*, 82, 726–735.

E. S. Gilbert, L. Huang, A. Bouville, C. D. Berg, and E. Ron, 2010, Thyroid cancer rates and ^{131}I doses from Nevada atmospheric tests: An update, *Radiation Research*, 173, 659–664.

S. Glasstone and P. J. Dolan, 1977, *The Effects of Nuclear Weapons*, 3rd Ed., U.S. Government Printing Office, Washington, D.C., 653 pp.

P. S. Harris, S. L. Simon, and S. A Ibrahim, 2010, Urinary excretion of radionuclides from Marshallese exposed to fallout from the 1954 Bravo nuclear test, *Health Physics*, 99, 217–232.

F. O. Hoffman, A. I. Apostoaei, and B. A. Thomas, 2002, A perspective on public concerns about exposure to fallout from the production and testing of nuclear weapons, *Health Physics*, 82, 736–748.

S. A. Ibrahim, S. L. Simon, A. Bouville, D. Melo, and H. L. Beck, 2010, Alimentary tract absorption (f_1 values) for radionuclides in local and regional fallout from nuclear tests, *Health Physics*, 99, 233–251.

T. Imanaka, 2006, Casualties and radiation dosimetry of the atomic bombings on Hiroshima and Nagasaki, in *Radiation Risk Estimates in Normal and Emergency Situations* (A. A. Cigna and M. Durante, eds.), Springer, New York, 149–156.

International Atomic Energy Agency (IAEA), 2002, *IAEA Safeguards Glossary, 2001 Edition*, Vienna, Austria, 218 pp, http://www-pub.iaea.org/MTCD/publications/PDF/nvs-3-cd/PDF/NVS3_prn.pdf.

A. W. Klement, 1959, A review of potential radionuclides produced in weapons detonations, *Report WASH-1024*, U.S. Atomic Energy Commission, Washington, D.C., 102 pp, http://www.osti.gov/bridge/servlets/purl/4230448-WNjgSt/4230448.pdf.

A. W. Klement, 1965, Radioactive fallout phenomena and mechanisms, *Health Physics*, 11, 1265–1274.

C. E. Land, Z. Zhumadilov, B. I. Gusev, M. H. Hartshorne, P. W. Wiest, P. W. Woodward, L. A. Crooks, N. K. Luckyanov, C. M. Fillmore, Z. Carr, G. Abisheva, H. L. Beck, A. Bouville, J. Langer, R. Weinstock, K. I. Gordeev, S. Shinkarev, and S. L. Simon, 2008, Ultrasound-detected thyroid nodule prevalence and radiation dose from fallout, *Radiation Research*, 169, 373–383.

C. E. Land, A. Bouville, I. Apostoaei, and S. L. Simon, 2010, Projected lifetime cancer risks from exposure to regional radioactive fallout in the Marshall Islands, *Health Physics*, 99, 201–215.

M. P. Little, D. G. Hoel, J. Molitor, J. D. Boice, Jr., R. Wakeford, and C. R. Muirhead, 2008, New models for evaluation of radiation-induced lifetime cancer risk and its uncertainty employed in the UNSCEAR 2006 Report, *Radiation Research*, 169, 660–676.

M. May, J. Davis, and R. Jeanloz, 2006, Preparing for the worst, *Nature*, 443, 907–908.

L. Machta, R. J. List, and L. F. Hubert, 1956, World-wide travel of atomic debris, *Science*, 124, 474–477.

Y. Morino, T. Ohara, and M. Nishizawa, 2011, Atmospheric behavior, deposition, and budget of radioactive materials from the Fukushima Daiichi nuclear power plant in March 2011, *Geophysical Research Letters*, 38, L00G11, doi:10.1029/2011GL048689.

B. E. Moroz, H. L. Beck, A. Bouville, and S. L. Simon, 2010, Predictions of dispersion and deposition of fallout from nuclear testing using the NOAA-HYSPLIT meteorological model, *Health Physics*, 99, 252–269.

National Research Council (NRC), 1999, *Exposure of the American People to Iodine-131 from Nevada Nuclear-Bomb Tests: Review of the National Cancer Institute Report and Public Health Implications*, National Academies Press, Washington, D.C., 288 pp.

———, 2005a, *Effects of Nuclear Earth-Penetrator and Other Weapons*, National Academies Press, Washington, D.C., 150 pp.

———, 2005b, *Assessment of the Scientific Information for the Radiation Exposure Screening and Education Program*, National Academies Press, Washington, D.C., 430 pp.

———, 2006, *Health Risks from Exposure to Low Levels of Ionizing Radiation: BEIR VII – Phase 2*, National Academies Press, Washington, D.C., 424 pp.

Organization for Economic Co-operation and Development (OECD), 2002, *Chernobyl, Assessment of Radiological and Health Impacts*, Paper 230, OECD Nuclear Energy Agency, Issy-les-Moulineaux, France, 155 pp.

M. Peplow, 2006, Counting the dead, *Nature*, 440, 982–983.

A. Robock, L. Oman, G. L. Stenchikov, O. B. Toon, C. Bardeen, and R. P. Turco, 2007, Climatic consequences of regional nuclear conflicts, *Atmospheric Chemistry and Physics*, 7, 2003–2012.

S. L. Simon and A. Bouville, 2002, Radiation doses to local populations near nuclear weapons test sites worldwide, *Health Physics*, 82, 706–725.

S. L. Simon, A. Bouville, and H. L. Beck, 2004, The geographic distribution of radionuclide deposition across the continental U.S. from atmospheric nuclear testing, *Journal of Environmental Radioactivity*, 74, 91–105.

S. L. Simon, A. Bouville, and C. E. Land, 2006, Fallout from nuclear weapons tests and cancer risks, *American Scientist*, 94, 48–57.

S. L. Simon, A. Bouville, C. E. Land, and H. L. Beck, 2010a, Radiation doses and cancer risks in the Marshall Islands associated with exposure to radioactive fallout from Bikini and Enewetak nuclear weapons tests: Summary, *Health Physics*, 99, 105–123.

S. L. Simon, A. Bouville, D. Melo, H. L. Beck, and R. M. Weinstock, 2010b, Acute and chronic intakes of fallout radionuclides by Marshallese from nuclear weapons testing at Bikini and Enewetak and related internal radiation doses, *Health Physics*, 99, 157–200.

A. Stohl, P. Seibert, G. Wotawa, D. Arnold, J. F. Burkhart, S. Eckhardt, C. Tapia, A. Vargas, and T. J. Yasunari, 2011, Xenon-133 and caesium-137 releases into the atmosphere from the Fukushima Dai-ichi nuclear power plant: determination of the source term, atmospheric dispersion, and deposition, *Atmospheric Chemistry and Physics—Discussions*, 11, 28319–28394.

C. Stubbs, 2011, The interplay between civilian and military nuclear risk assessment, and sobering lessons from Fukushima and the Space Shuttle, chapter 3, this volume.

O. B. Toon, R. P. Turco, A. Robock, C. Bardeen, L. Oman, and G. L. Stenchikov, 2007, Atmospheric effects and societal consequences of regional scale nuclear conflicts and acts of individual nuclear terrorism, *Atmospheric Chemistry and Physics*, 7, 1973–2002.

United Nations Scientific Committee on the Effects of Atomic Radiation (UNSCEAR), 2000, Exposures and Effects of the Chernobyl Accident, in *Sources and Effects of Ionizing Radiation*, Vol. 2, Annex J, 451–566, http://www.unscear.org/unscear/en/chernobyl.html.

———, 2008, Health effects due to radiation from the Chernobyl accident, in *Sources and Effects of Ionizing Radiation*, Vol. 2, Annex D, 45–219, http://www.unscear.org/unscear/en/chernobyl.html.

———, 2010, *Report on the Effects of Atomic Radiation, Summary of Low-Dose Radiation Effects on Health*, 14 pp., http://www.unscear.org/unscear/en/chernobyl.html.

Notes

1. This could occur with an improvised nuclear device or a nuclear weapon stolen or otherwise obtained from a military stockpile.

2. See appendix for additional background information.

3. The fallout-free height of burst identifies a detonation point at high enough altitude that, for a given yield, the fireball does not reach the ground (Glasstone and Dolan, 1977, p. 71; NRC, 2005a). Fallout from the main cloud is therefore lofted upward and dispersed (where it has more time to decay), rather than reaching ground level, unless scavenged by precipitation. This is in contrast to a ground or subsurface burst, which incorporates considerable amounts of soil, rock, and other surface material into the fireball, making for larger amounts of radioactive material (due to neutron activation) and increased likelihood of radioactive dispersal at ground level. Neutron activation is a process by which a fraction of the (originally) non-radioactive material at ground level is made radioactive by the burst of neutrons released from the nuclear explosion.

4. Glasstone and Dolan (1977, pp. 416–418) refer to "scavenging" of the radioactive cloud in describing precipitation (rain, snow, etc.) depositing radioactive debris at ground level.

5. The unit of dose from ionizing radiation is 1 gray (Gy) = 1 Joule/kilogram = 1 J/kg (= 100 rad, in older units) of absorbed radiation; 1 mGy = 0.001 Gy = 0.1 rad. The dose-equivalent unit is 1 sievert (Sv), which has the same units of 1 J/kg (= 100 rem, in older units): it is intended to take into account differences in biological effects of different forms of radiation (different particles and energies) and different organs' biological sensitivity to ionizing radiation; 1 mSv = 0.001 Sv = 0.1 rem. For our purposes, 1 Sv is the same as the absorbed

dose of 1 Gy. More generally, equating the two units can underestimate the biological effects of ionizing radiation for a given dose in Gy.

6. This residual radiation was due to neutron activation of surface material (Imanaka, 2006). See note 3.

7. Similarly, measurement of radioactivity in Troy, New York, after the April 25, 1953, SIMON tower shot in Nevada revealed the great distances (~ 3600 km) which fallout from a nuclear explosion can travel, and then be rained out onto the ground (Clark, 1954; Machta, et al., 1956; DHHS, 2000, appendix E; Simon, et al., 2006). Clark (1954) estimates a 10-week accumulated dose of ~ 0.5 mGy in Troy (dose rate one day after deposition being nearly thirty times background). See also Hoffman, et al. (2002).

8. The decay rate of early (local) fallout from a nuclear weapon can be more than five times faster than for the debris released from a nuclear power plant (Glasstone and Dolan, 1977; Eisenbud and Gesell, 1997). See appendix for additional information.

9. This is Glasstone and Dolan's (1977) reference dose rate for early (local) fallout assuming a wind speed of 24 km/h = 15 miles per hour. One ton of TNT equivalent yield corresponds to an energy release of 4.18 trillion joules (4.18 TJ); 1 kiloton (kt) and 1 megaton (Mt) are equal to the explosive yield from 1000 tons and 1 million tons of TNT, respectively. Explosion-yield values refer exclusively to fission yield throughout this article.

10. http://www.stuk.fi/sateilyvaara/en_GB/esim_annos.

11. For simplicity, all explosions are assumed to occur at one location, whereas a more realistic spatial dispersion of detonation sites would expand the width and decrease the average dose levels of the fallout-affected region.

12. The values given here are for accumulated doses (mGy or mSv), whereas the numbers in an earlier paragraph were for hourly dose rates (mGy/h or mSv/h).

13. The Nevada National Security Site (N2S2), as of 2010.

14. 1990 Radiation Exposure Compensation Act, http://www.justice.gov/civil/common/reca.html.

15. Studies of fallout around other nuclear-explosion test sites provide important complementary information (e.g., Land, et al., 2008, 2010; Simon, et al., 2010a,b; Beck, et al., 2010; Bouville, et al., 2010; Harris, et al., 2010; Ibrahim, et al., 2010; Moroz, et al., 2010; de Vathaire, et al., 2010).

16. The values quoted here for adults' bone marrow and children's thyroids span the range of doses reported for fallout-induced health effects (low to high values, respectively), and the outer values of distances quoted indicate ranges

to locations of outliers in the patterns of fallout (Figure 4.1). Some of the key isotopes considered include Iodine^{131} and Cesium^{137} (DHHS, 2005, p. 26).

17. The estimated 95-percent uncertainty limits are given by the values in parentheses: that is, the actual value is in this case estimated with 95-percent confidence to lie between 26,000 and 237,000 person-Gy. The exposure estimates include taking into account shielding experienced by individuals through their daily work and life cycle (DHHS, 2005: chapter 3 and appendix E); such estimates of shielding are necessarily approximate, and may not apply outside the United States.

18. In order to derive the dose that would have been produced by a total fission yield of one megaton of ground bursts, I have multiplied the dose by 1.33 to account for the reduced fallout from air and balloon bursts (Beck and Bennet, 2002; DHHS, 2005: see Figure 8 and associated text in appendix E).

19. Tables 12.5B and 12.8 of BEIR-VII (NRC, 2006) give separate estimates of 0.051/Gy and 0.061/Gy for lifetime attributable risk (LAR) mortality due to solid cancers, and 0.007/Gy for LAR mortality due to leukemia (see also Little, et al., 2008, and UNSCEAR, 2010). I have averaged and rounded off to get an approximate value of 0.06/Gy for LAR mortality due to external exposure.

20. Pages 100 and 107 of NRC (1999). Only those exposed as young children (under 10 years old) appear to be susceptible.

21. The U.S. population at the time was 163 million (see DHHS, 2005: Table 3.14), of which roughly 150 million can be considered to have been downwind.

22. These values are based on older LAR mortality factors than in BEIR-VII (NRC, 2006), and would be slightly larger using the more recent numbers (see note 19).

23. The ~5,600 additional solid-cancer deaths and 750 additional leukemia fatalities expected as a result of *external* exposure to radioactive fallout from one megaton ground-level explosions (see footnotes 17–18) correspond to 0.01 percent of 60 million and 0.05 percent of 1.5 million of *total* cancer and leukemia deaths, respectively, projected for the same population. Additional deaths due to *internal* exposure to radioactive fallout are expected to represent a larger fraction of mortality for the same population.

24. Concern is compounded by mistrust in the reliability or completeness of information given to the public regarding radioactive fallout, a case in point being the faulty initial reports of radioactive releases from the Chernobyl accident (Table 4A.1) followed by the controversial summary in UNSCEAR 2000 (p. 486–488) of the eventual consequences of this radioactive fallout for

public health (e.g., Garwin and Charpak, 2001, pp. 186–195; Peplow, 2006; see also Anspaugh, et al., 1988; OECD, 2002; Cardis, et al., 2006; and UNSCEAR, 2008).

25. National and international reaction to the 2,977 deaths from the September 11, 2001, attacks on the United States gives a crude sense of this distinction, depending on context: these deaths amount to less than five weeks of highway fatalities, or 1 in 105,000 of the U.S. population.

26. These numbers are subject to many uncertainties, not least due to the estimated effectiveness of shielding (see note 17: it is possible that sheltering in different parts of the world may be either more or less effective than is thought to have been the case in the United States).

27. The unit of activity, 1 becquerel (Bq) = 1 disintegration per second (= 2.7×10^{-11} curie (Ci) in older units), is a combined measure of both the abundance of a radioactive substance and its rate of decay. This describes the intensity of radioactivity that could, for example, affect health. It is commonly used in larger magnitudes: 1 PBq = 1,000 trillion Bq = 27,000 Ci. The relationship between measured activities and inferred doses (Gy) or dose-equivalents (Sv) of ionizing radiation is discussed in DHHS (2005) and references therein (see note 5).

28. Put another way, because one kilogram of fission in a bomb contributes seventeen kilotons of energy release, a typical nuclear reactor producing 1,000 megawatts (1 GW) of electric power—which fissions 1,000 kilograms of material per year—creates seventeen megatons worth of nuclear fission products per year (i.e., an amount corresponding to forty-six kilotons of nuclear fission explosions per day).

29. The United Nations' International Atomic Energy Agency (IAEA) defines a "significant quantity" of plutonium (Pu) as eight kilograms, and of highly enriched uranium (HEU: $^{235}U \geq 20$ percent) as twenty-five kilograms (IAEA, 2002, Table II: p. 27). Note that a large fraction of the nuclear explosive material is typically not consumed in a nuclear explosion (e.g., May, et al., 2006). Decay rates are proportional to the reciprocal of half-lives (i.e., long half-life implies slow decay rates, and vice versa), and the half-lives of ^{235}U and ^{239}Pu are 0.7 billion years and 24,000 years, respectively.

30. Other radioactive materials having much faster decay rates (short half-lives)—such as tritium—may also be present, but are in such small abundances as to be relatively inconsequential.

Naval Nuclear Power as a Model for Civilian Applications

5

DREW DEWALT

For over half a century, the United States Navy has operated nuclear reactors without a major safety incident. Therefore, it serves as an exemplar for safe operations involving highly complex systems. The purpose of this paper is to discuss the guiding principles that contributed to the Navy's nuclear safety record while acknowledging the inherent differences between military nuclear power and civilian applications. Further, suggestions are offered to better align civilian operations with military operations in regard to safety.

Introduction

To begin, it must be acknowledged that the safety record of the Navy's nuclear power program is not without blemish. There have been minor incidents involving inadvertent release of low-level radioactivity and radioactive material. If a *major* incident is defined as one in which there is a release of high levels of radioactivity with human health risks and lasting effects on the environment, then the claim can be made that the Navy has an impressive safety record. In this regard, civilian nuclear enterprise can benefit greatly by adopting many of the principles that have guided the Navy's nuclear power program. This is especially true when you consider that the main causes of the more infamous civilian nuclear accidents have been violations of these guiding principles—notably Three Mile Island (inadequately trained operators); Fukushima (lack of conservatism of design, living with deficiencies); and Chernobyl (lack of conservatism of design, inadequately trained operators, and non-compliance with approved operating procedures). For the civilian nuclear industry to survive and prosper in a post-Fukushima world, the industry must take aggressive steps to operate even more safely, and thus promote public confidence in nuclear power as a safe and viable alternative to electricity generation using fossil fuels.

The Principles[1]

In the wake of the Three Mile Island accident, Admiral Hyman G. Rickover outlined more than a dozen key principles to ensure the safety of naval nuclear power. This paper focuses on the most important of these principles, which are also most readily applicable to civilian nuclear power. Like most industries involving complex systems, the nuclear industry often considers that technological advances and improved design alone can provide long-term safety and operational capability. However, the human factors involved in operations are arguably more important to ensuring safety. The principles identified below are the most significant for securing the safety of all aspects of operations. Unfortunately, they also seem to be the most neglected in civilian enterprises.

1. *Strong central technical control.* Design, manufacturing, assembly, testing, operations, maintenance, and selection and training of personnel should not be separated. All are so interrelated that the entire nuclear operation requires close technical coordination and direction. This can be reasonably achieved only by having strong

central technical control vested with both the responsibility and the authority to maintain the integrity of the entire program. Implicit in this principle is the raising of issues and potential problems to the central body for review, for a decision regarding the proper corrective action, and for the conducting of frequent, thorough, detailed audits of all aspects of a nuclear program.

2. *Technical competence.* All individuals who make decisions in design, operation, maintenance, training, and acquisition must understand the technology involved. Expectations for knowledge and training cannot be limited to the basic tasks of each individual's specific job. Instead, everyone involved in the enterprise must understand the overarching technology in order to comprehend and properly appreciate the implications of decisions and actions.
3. *Conservatism of design.* The reactor design and associated power plant must, from the beginning, allow for every uncertainty and inaccuracy in the available knowledge. Additional design features and safety systems do not and cannot compensate for an overall lack of conservatism of design. This principle was most glaringly violated at Chernobyl, where the design incorporated both a positive temperature feedback and a positive void coefficient. This basically means that as temperature rises or more steam voids are created, there is a positive feedback that causes reactor power to increase as well. For example, more conservative reactors would be designed to have negative temperature coefficients such that the rising temperature associated with rising reactor power has a mitigating effect on the power transient as opposed to an accelerating effect that a positive coefficient would have. This principle of conservatism of design has significant potential of being violated in newer reactor designs of countries with less experience in operating nuclear reactors.
4. *Compliance with detailed operating procedures.* Nuclear power plant operations must be accompanied by detailed, written procedures for all standard evolutions and potential casualties. Any deviations from procedures cannot be made locally, but instead must require

approval by the central technical body. In addition, formal documentation must be used in all parts of the nuclear program to include normal operations, maintenance, identification of potential problems, and casualty situations.

5. *Willingness to face the facts.* This principle summarizes the culture that must be developed to safely sustain a nuclear enterprise. Everyone, from decision-makers and operators all the way down to the janitors, must be encouraged to bring questions and possible problems to the attention of plant managers, who in turn should be empowered to further raise those same concerns to upper management. The rule should be: when in doubt, notify a supervisor. In this way, the operators of a nuclear power plant do not tolerate deficiencies. This is line management, top to bottom, such that everyone is responsible for safety including floor workers, plant managers, and the companies' top executives—responsibility and accountability up and down the line. Safety cannot be ensured by only one part of an organization; it has to be integral to all parts of a nuclear enterprise. Managers must also be held accountable for long-term performance and safety, and associated corporate offices must be willing to make necessary changes despite significant costs and delays in the operating schedule.

Difficulties of Civilian Nuclear Power

Because of the inherent differences between civilian and military nuclear programs, it is more difficult for civilian enterprises to operate as the Navy's nuclear power program operates. Most evidently, civilian utilities need to operate at a profit while providing reliable power at minimum cost. This alone points to several influences that can run counter to overall safe operations. Many times, conservative reactor design suggests adding multiple barrier systems and redundancies to protect against equipment failure, human error, and other adverse events. These systems themselves are costly and can also prevent a reactor from operating near its design limits such that output per time in operation remains lower than what is otherwise achievable. Also, there is pressure to serve customers and meet consumer demand. Lastly, the need to operate at a profit can often mean competing with other forms of energy generation on a cost basis.

Other difficulties include the existence of multiple different reactor designs. This can result in the need to keep much of the technical information private so as to maintain an advantage over competitors. Also, more than the military, civilian nuclear operators must deal with public hesitancy surrounding nuclear power, due to past nuclear accidents and a "not in my backyard" mentality. Despite these difficulties, the civilian nuclear industry must operate more safely than it currently does. The industry most likely would not survive another Fukushima-like accident.

Suggestions to Improve Safety

The first step to improve the safety of civilian nuclear operations is to have a central oversight body with authority to manage all technical aspects of operations and to conduct independent audits irrespective of schedules and economic concerns. This central body must be completely independent of all utilities and vested with the authority to shut down a reactor's operations any time its continued operation is deemed to be unsafe. This central organization should also approve all reactor designs, procedures, and changes to established procedures. In the United States, the Nuclear Regulatory Commission (NRC) is entrusted with this mission, but in its current form the NRC is only nominally independent from private industry interests. Industry lobbyists are able to affect regulation through political channels influencing the NRC's budget, and a "revolving door" exists between private industry players and nuclear regulators. As a result, there is evidence that the NRC's regulations are not strictly enforced in all instances, and deficiencies that are identified are allowed to linger (overview provided in source 7). Therefore, NRC independence must be ensured and its regulations enforced, such that safety behaviors in the industry improve.

The second recommendation would be to push for standardization in the nuclear industry. This would involve standardization of equipment, materials, and procedures. This would drive down costs in the industry while promoting learning and sharing of best practices. Such standardization would also allow the United States to better attract, train, and retain talented personnel for the nuclear industry. Standardization would make it easier for the NRC to audit and regulate nuclear industry operations. This recommendation acknowledges that some aspects of each company's operations such as specialized equipment will need to remain proprietary, but standardization to the furthest extent possible in a competitive industry would be highly beneficial.

The third recommendation is to enforce a system that holds top management personally responsible for reactor plant operations. This implies that top management should be technically competent, and technical decisions should not be subordinate to accountants, lawyers, and professional managers. If a reactor plant has significant safety issues, the repercussions should extend past the local plant manager to the company's executives. It is the responsibility of top management to build and maintain a culture where issues are identified, elevated, and corrected such that they never become serious safety issues. The responsibility and accountability for safety up and down the line is the backbone for safe operations. This level of accountability encourages organizations to maintain a long-term focus on safety that will not be sacrificed for short-term output and revenue goals.

In Closing

Despite public concerns and the political expediency of criticizing the nuclear industry, the threat of discontinuing the use of civilian nuclear power is currently unrealistic. Nuclear power provides over 20 percent of the United States' installed base-load electricity generation capacity, so such a move would cripple the U.S. economy and push the country further from energy independence. That being said, if a Fukushima-like incident were to occur on American soil (or even a smaller incident closer to that of Three Mile Island), the chance of discontinuing civilian nuclear power operations would rise dramatically. This is why it is imperative to immediately improve the safety of operations. Advances in technology and reactor design already help the industry operate more safely but, as previously discussed, such improvements must be accompanied by enhanced human systems to ensure operational safety.

References

1. Rickover, Admiral H. G. "Differences Between Naval Reactor and Commercial Nuclear Plants." Comments subsequent to the accident at Three Mile Island (August 1979).
2. Perrow, Charles. "Fukushima and the Inevitability of Accidents." *Bulletin of Atomic Scientists,* Issue 67 (2011): p. 44–52.
3. Kaufmann, Daniel. "Preventing Nuclear Meltdown: Assessing Regulatory Failure in Japan and the United States." *Brookings* (April 1, 2011).
4. Union of Concerned Scientists. "Nuclear Power and Global Warming." Position paper (March 2007).
5. United States Nuclear Regulatory Commission website, http://www.nrc.gov/. See especially, "Commission Responses to GAO Reports."
6. U.S. NRC, "Performance and Accountability Report: Fiscal Year 2011," http://www.nrc.gov/reading-rm/doc-collections/nuregs/staff/sr1542/v17.
7. Zeller Jr., Tom. "Nuclear Agency Is Criticized as Too Close to Its Industry." *The New York Times* (May 7, 2011).

Session II

Nuclear Reactor Safety

Lessons Learned of "Lessons Learned": Evolution in Nuclear Power Safety and Operations

6

EDWARD BLANDFORD AND MICHAEL MAY

Executive Summary

In this paper, we survey briefly the lessons that emerged from the three major accidents in nuclear power history—the Three Mile Island (TMI) accident in the United States in 1979, the Chernobyl accident in Ukraine in 1986, and the Fukushima accident in Japan in 2011—as well as from a few less important accidents. We then note which of these lessons were learned in the sense that measures to prevent recurrence of the accidents were adopted. We conclude the paper with a few observations bearing on possible future action that emerge from our survey.

Here are the major general observations from our survey:

1. In terms of fatalities and effects on health and environment, nuclear power has been better than—or at least as good as—the other ways of generating electricity. Nevertheless there is no way to ensure complete safety in the nuclear industry or anywhere else. Continuing to learn from every opportunity is therefore essential but has occurred spottily, especially across national boundaries. "Safety is hard work. It must be embedded in the management and cultural practices of both operators and regulators; it is an obligation that demands constant attention."[1] This obligation has not always been met.

2. All three of the major nuclear power accidents (TMI, Chernobyl, and Fukushima) as well as several of the less-well-known close calls had precursors in previous incidents, although often not in the same country. The "lessons learned" reviews that have followed most usually made specific useful points. Some of those were implemented—the lessons were learned—but often they were not. Not surprisingly, implementation steps that translated into more efficient operations, such as better, more standardized operating procedures, were carried out more often than steps that required immediate expenditures to avoid uncertain disaster, such as better defenses against flooding. Further analysis may find other, less obvious correlations.

3. A regulating agency with appropriate power and strong technical competence, well-staffed and funded and independent of its licensees,[2] is a necessary—though not sufficient—requirement for safety and in particular for the formulation and implementation of lessons learned. Regulator capture by licensees through either political or administrative processes has been a problem in several countries. Recent moves (e.g., in India) to remove the regulating agency from the administrative structure of the operating and promoting agency are a step toward greater safety. Beyond an effective regulator, however, a culture of safety must be adopted by all operating

entities. For this, the tangible benefits of a safety culture must become clear to them. Regulators must also encourage the identification and reporting of problems, to enable effective implementation of corrective action programs. Effectively ensuring nuclear safety and, for that matter, security at nuclear sites is not a static matter of setting forth regulations to meet known problems. Rather, it is a continuing and dynamic set of interactions involving regulators, licensees, and other stakeholders, none of which is independent of the others.

4. An example of a well-balanced combination of transparency and privacy is INPO (the Institute of Nuclear Power Operations), which was created in the wake of TMI. Funded and supported by the U.S. nuclear power industry, INPO provides an ongoing process for learning lessons in the operations area: operator ratings at the various plants remain private but results in terms of operating procedures and consequences are public. Because of differing laws, policies, and priorities, however, it will be difficult to extend the concept to the international nuclear power industry despite the fact that what happens in one country usually affects the future in other countries. In addition, many lessons that do not concern operations must be learned. International cooperation should be broadened beyond participation in INPO and the World Association of Nuclear Operators (WANO) to include, e.g., the Electric Power Research Institute (EPRI).[3]

5. In the United States and in some other countries, public fear of radioactivity and the ensuing interventions of often well-informed organizations have been a spur to learning from experience. On the other hand, unswerving ideologically based political opposition has served to decrease transparency and mutual cooperation.

6. Because so much of the cost of nuclear power is incurred before the first kilowatt-hour is generated, the financial backers, including private and government insurers and guarantors, have in theory considerable leverage over the industry, as does any entity

that can delay construction and operations, such as regulators and interveners.

7. There can be a tendency to focus the "lessons learned" effort primarily on system failures, sometimes marginalizing system successes. Lessons can also be learned from successes. Severe reactor accidents are extremely rare and every effort should be taken to abstract key engineered or organizational successes.

If we look at the conclusions drawn from lessons learned and not learned and ask what their application is to the future and, in particular, to how best to improve nuclear power safety worldwide, we come to another set of observations.

1. Modern reactors (classified as Generation III and III+) use safer designs and can be operated more safely than the ones that have caused major accidents. But it is not clear at present how many of the safest designs will be built. Currently there are over sixty new reactors under construction and hundreds more in the planning stage. The majority of those under construction are Generation II designs with enhancements over plants currently operating. However, the first sets of Generation III and III+ designs are now being built and many reactors in the planning stage will incorporate the improved variety.
2. The Fukushima accident was caused by a "one-in-a-thousand-years event" plus some admittedly faulty design and siting. The precursor incident at the Blayais nuclear plant in France had also been viewed as a "one-in-a-thousand-years event." Nevertheless, looked at on a worldwide basis and considering the proposed lifetimes of modern reactors, there is a valid statistical basis for taking such events into account and spending some money to prevent or alleviate their consequences.
3. Mechanisms to facilitate and, where needed, enforce mutual learning among countries is not as effective today as it is within countries and may not be adequate to prevent avoidable disasters. Information-sharing, import-export agreements based on safety

standards, agreements to facilitate cooperation among regulatory authorities, and the participation of financial interests such as investors and insurers may all have a role in improving mutual learning among different states.

4. Improved cooperation will rest most securely on lasting shared economic interest among vendors, owners-operators, government regulators, and the public. At the same time, the international nuclear power and nuclear fuel cycle markets will become, if anything, more competitive than they have been. New users with no operating experience and no regulatory experience are entering the market, e.g., the United Arab Emirates (UAE). Therefore, without considerable government attention and cooperation, the nuclear power industry may not necessarily become safer, although it has the potential to do so from a purely technical point of view by adopting the more advanced Generation III and III+ passive designs.
5. Any plan to deal with emergencies must include an incident command structure with clear lines of communication and well-defined areas of responsibility, including the responsibility to provide timely information to the actors involved and to the public. This plan must include all relevant actors, from top political authorities to the regulators and management structure of the licensee and on to local operators and responders at the scene of the emergency. Preliminary reviews of the Fukushima accident have noted the need for improvement in that particular case, but both Chernobyl and, to a lesser extent, TMI demonstrated the need for improvement in this area.

Background

Safety issues associated with nuclear technology first arose during the Manhattan Project, during the establishment of the U.S. nuclear weapons program. In 1942, the DuPont company agreed to be the prime contractor responsible for construction of the plutonium production complex, starting initially at Oak Ridge, Tennessee, and ultimately being completed at the Hanford site in Washington state. Nuclear technology spanning the fuel cycle from enrichment all the way to chemical separation was just entering its infancy at a remarkable pace, with large material inventory demands and little margin for error.

In fact, it was DuPont chemical engineers working on the B-Reactor at Hanford who first formally introduced reactor system hierarchy and the "defense-in-depth" concept into reactor design and construction.[4] The B-Reactor was the first large-scale reactor built following the successful demonstration of the technology at Oak Ridge with the X-10 pilot reactor. Due to the unfamiliarity of the technology, the DuPont engineers relied upon their fundamental understanding of industrial chemical plants and implemented several layers of independent "barriers" between the site workers and the hazardous radioactive source. Additionally, the concepts of redundancy and diversity in engineered safety systems were first formalized into the reactor design process.

Out of the weapons program emerged a commercial nuclear industry that has undergone many transformations over the last fifty years. In this paper, we focus on the way the organizations responsible for operating and regulating this industry have learned from operational experience, their own and that of others.[5] Throughout this history there have been a range of reactor events differing in severity. Many of these events have been deconstructed and better understood through root-cause investigations yielding a set of "lessons learned." Our focus in this paper is to examine these sets further and develop insights about how the industry and other stakeholders collectively learn from accident experience. Following the three major commercial reactor accidents—TMI, Chernobyl, and Fukushima-Daiichi—the lessons-learned process was carried out in public and scrutinized by the media. However, there have also been less severe incidents and operational anomalies that have received much less attention but have, in some cases, provided invaluable learning experience. What lessons were learned as compared with lessons that should have been learned and were not? How can this experience inform the future so it can be better than the past?

Key Stakeholders Involved

The key organizations that are responsible for industry learning include the regulatory and other relevant government authorities, licensees and their shareholders, industry organizations, media, and citizen groups. There is a special need for nuclear installations to demonstrate and maintain higher standards of safety than the utility fossil industry norm, given the potential for severe accidents at some of the installations and the public apprehension over things nuclear. There is therefore a need for all stakeholders to make full use of the lessons-learned process. Additionally, regulatory bodies *and* licensees both have to learn from serious accidents. This requires regulatory independence from politics and transparency, among other factors; we look at these questions as they affect stakeholder groups.

Historically, one of the challenges with establishing effective regulatory bodies is ensuring the complete separation of the organizations responsible for advancing and implementing the technology from those charged with regulating it, as well as insulating the regulators from political pressures as much as possible.[6] The two types of agency were originally combined due to heavy federal involvement with introducing the technology commercially. This split occurred for different motivations and with different effectiveness in different countries. The United States split the Atomic Energy Commission into the Nuclear Regulatory Commission and the Energy Research and Development Administration in 1974 due primarily to political and confidence reasons. Other countries, such as India and, most recently, Japan, have started along the same road following Fukushima-Daiichi.

Private organizations such as INPO and WANO perform important functions and are discussed later in the context of learning from accidents. Members of the public, through non-governmental organizations (NGOs) and the intervention process, have also played roles in the lessons-learned process, roles that can vary internationally.

Evaluating Off-normal Operation

In order to combine into an effective system, both licensee and regulator must constantly learn from *all* modes of operation. Success and failure in nuclear operation are continuums and must be evaluated with equal scrutiny. Success does not only require meeting regulatory requirements and maintaining high capacity factors. It is a dynamic process that includes learning. Conversely, failure in plant operation can include routine maintenance all the way up to catastrophic failure. All of this enters into a dynamic process of improvement. Success and failure can be measured in such variables as economic, health, and environmental impacts.

In this paper, we discuss events that have occurred since the inception of the commercial nuclear industry. For the purpose of this paper, we will loosely follow the qualitative and therefore somewhat subjective International Nuclear and Radiological Event Scale (INES) of the International Atomic Energy Agency (IAEA), according to which events are rated from "operational anomalies" through "incidents" and finally all the way to "severe accidents." The INES scale considers the impact on people and the environment, radiological barriers and control, and defense-in-depth. Each event considered in this paper is shown by the level on the INES scale in Figure 6.1. The role of "precursor events" that can lead to accidents will be seen to be essential: severe accidents are often the results of earlier anomalies and incidents. Successful identification of these precursors requires initiative, awareness, and operational experience.

General Assumptions

To focus the discussion, we have made some initial assumptions surrounding the relevant background. Some of these assumptions are fleshed out further in additional papers written for this meeting. These assumptions include the following statements:

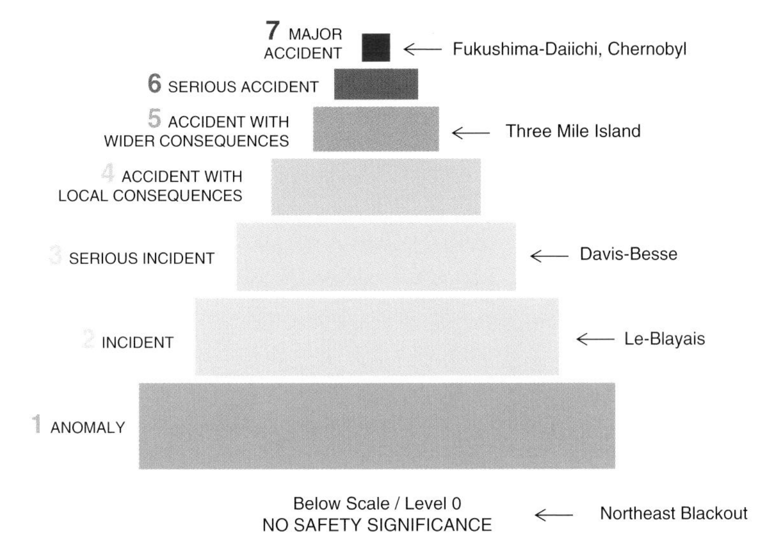

Figure 6.1 This figure indicates where each event considered in this paper lies on the IAEA International Nuclear and Radiological Event Scale (INES).
Source: Adapted from IAEA, "The International Nuclear and Radiological Event Scale," http://www.iaea.org/Publications/Factsheets/English/ines.pdf.

- Risk acceptance varies widely across the world. This paper is normative in the sense that it represents a local perspective and is not globally representative.
- It is critical that we differentiate reactor technology and plant operations as they involve fundamentally different organizations and this relationship varies widely across the world.[7]
- Initiating events can be broadly classified as "internal" or "external." Internal events are typically caused by combinations of hardware failures and human errors. External events can be malicious (e.g., terrorist attack) or natural hazards such as earthquakes or tornados. Some events that can be internally or externally initiated, like fire and flooding, are often classified as external events.

- The term "near miss," while not quantified, is used when damaged or deteriorating equipment, human error, or some other factor internal to the state of the reactor or its operation increases the risk of core damage sufficiently to cause the NRC to send out an inspection team (IT).
- Lessons are learned from both success and failure. Severe reactor accidents are extremely rare and every effort should be taken to abstract key engineered or organizational successes.

Key Reactor Accidents, Incidents, and Anomalies

In this section, we review the lessons-learned experience following a range of reactor events. We first focus on major reactor accidents where the lessons-learned process played out in the public domain and many stakeholders were involved in the process. The location of the reactor accident as well as the reactor technology heavily affects the lessons-learned experience. We then review some less severe reactor incidents and anomalies, which can be of equal interest as they often show the conditions that can lead to more serious accidents.

Reactor Accidents

In each of the following reactor accidents, key organizations such as IAEA, state regulatory authorities, licensee organizations, and independent commissions initiated formal review processes. Each accident will be discussed within the context of the type of initiating event, the major contributor(s) to failure, and the extent of hazard consequence. In the case of Three Mile Island and Chernobyl, the initiating events that caused the accident were *internal* events and were exacerbated by human errors. In the case of Fukushima-Daiichi, the initiating event that caused the reactor accident was an *external* event in the form of an earthquake and subsequent tsunami. A common response to nuclear accidents outside the country where the accident occurred is a) we don't build our reactors that way, b) we don't operate them that way, and/or c) we understand the governing phenomenology. We keep these three perspectives in mind in the following discussion.

Three Mile Island

This event occurred on March 28, 1979, near Harrisburg, Pennsylvania, when a cooling malfunction and human errors caused part of the core to melt in TMI unit 2. The Three Mile Island Nuclear Generating Station has two PWR units (pressurized water reactors), both Babcock & Wilcox designs. Unit 1 generates 800 megawatts of electricity (MWe) and was commissioned in 1974; unit 2 is slightly larger at 900 MWe and began operation in 1978. The accident was initiated by a stuck open pilot-operated relief valve (PORV) in the primary system and was exacerbated by operator action following the initiating event.[8] Unit 2 was ultimately destroyed. Some fission product gas was released a couple of days after the accident, but not enough to cause any detectable dose to local residents above background levels. There were no injuries or adverse health effects. The TMI accident was caused by an internal initiating event and ultimately rated a Level 5 Accident with Wider Consequences on the INES scale. The accident sequence and post-accident forensics are discussed in much greater detail elsewhere.[9]

Following TMI, there were many efforts to conduct comprehensive studies and investigations of the reactor accident. Best known was the "Kemeny Commission" established by President Carter two weeks after the accident. The commission was chartered with the task of making a technical assessment of what occurred and a series of recommendations for the future based on these findings. The NRC created its own inquiry group headed by Washington, D.C., attorney Mitchell Rogovin. Following the review of the accident, the NRC established a Lessons Learned Task Force (LLTF) responsible for making suggested changes in fundamental aspects of basic plant safety policy.[10]

Lessons Learned from TMI

The Kemeny Commission made a series of recommendations concerning the NRC, the licensees, training, technical assessment, public health and safety, emergency planning, and the public's right to information.[11]

Four strong themes emerged from these recommendations and were broadly classified by Rees as management involvement, normative systems, learning from experience, and professionalism.[12] An important recommendation that does not quite fit under those categories is for better human factors engineering (HFE), i.e., the engineering that goes into operator-machine interactions. Early control rooms without such HFE greatly increased the burden on operators in an emergency.

- The first key lesson focused on the role of management in operating nuclear power plants. Prior to TMI, many in utility management viewed nuclear plants just as assets indistinguishable from fossil generation. Utility executives focused solely on plant output, leaving the challenging day-to-day operations of the plant to others in the company. This led to performance objectives that were sometimes inconsistent with the required and expected level of safety.[13]
- The second key lesson stems from the overly prescriptive nature of the regulatory structure. The normative landscape was made up of an impressive list of required documentation, rules, standards, and so on required to build and operate a nuclear plant. This led to unintended consequences. Among others, it left operators believing that their plants were completely safe as long as safety requirements had been met.[14]
- Third, the Kemeny Commission noted that previous operational experience elsewhere in the fleet had not been learned across the industry. In fact, learning from experience across the industry was viewed as a peripheral activity and not a necessary endeavor. The Kemeny Commission noted that the dominant hardware failure at TMI of a failed PORV had occurred eleven times earlier, but this operational experience with Babcock & Wilcox valves had not been shared.
- Finally, the Kemeny Commission noted that there was an overall lack of professionalism in the personnel who operated the plants. As a result, operating standards had suffered. It is interesting to note that the Kemeny Commission also called for a complete

> restructuring of the Nuclear Regulatory Commission and the abolishment of the five-member commission system. Not all of its recommendations were followed, however. The multi-member regulatory commission system is well entrenched in a number of areas in the United States, with members named by political authorities but, once confirmed, nominally independent of them. There is no clear consensus on what structure best assures such independence—or, rather, effectiveness—at managing an inherently interdependent process that involves many stakeholders.

The NRC conducted its own review of TMI and made a list of improvements in nuclear power plant operations, design, and regulation.[15] This review was performed independently, but it recognized many of the limitations identified by the Kemeny Commission. Some key important regulatory changes included the establishment of crucial equipment requirements and the identification of human performance as an integral component of a safe nuclear plant.

Traditionally, the NRC had left plant management strategies to the licensees and focused most of its effort on plant operations. This gap was largely remedied by the creation of INPO by the industry itself, just two weeks after the TMI accident. The creation of INPO is often cited as the major lesson learned from TMI, and for good reason. INPO confounds the expected norm of an organization that improves the safety and reliability of the nuclear industry; that is, INPO is a private regulatory bureaucracy that was set up by the industry and is funded directly by licensees. Following TMI, it was recognized that the nuclear navy has an extraordinary safety record and perhaps the commercial industry should learn more from the nuclear navy. This was evident by the fact that INPO's first CEO was retired Navy Admiral Eugene Wilkinson, who had served under Admiral Hyman Rickover. There are many reasons why INPO has been recognized as a successful organization; many of these are discussed in the following section. The fact that INPO interacts at three distinct hierarchical levels within the organization (the worker level, the manager level, and senior management and executive levels) makes it extraordinarily effective. Additionally, the naval influence can be seen in INPO's emphasis on establishing effective self-assessment and corrective action programs.

Chernobyl

In late April 1986, during an experimental systems test at unit 4 of the Chernobyl Plant about eighty miles north of Kiev in Ukraine, a sudden power surge caused the plant to become unstable. Attempts to initiate emergency cooling failed and resulted in more severe power excursions. The reactor pressure vessel ultimately failed and a massive explosion led to huge amounts of radioactive material being released into the environment. The accident, the worst in the history of nuclear power, was due largely to a reactor design that led to an unstable condition during the test as well as to operator errors, some due to lack of adequate information. The ultimate causes were complex and involved several of the reactor's design features, including its lack of secondary containment. Another factor was that under some conditions the more the coolant water boiled, the more power was generated and, again under some conditions, power generation also went up when the control rods designed to shut down the reaction were inserted. The reactor, a Soviet-designed RBMK that was originally deployed in several Soviet bloc countries, is now found only in Russia and is no longer being built. Several of the design features that led to the accident have been fixed. Much more on the accident can be found in a number of publicly available references covering the sequence of events, the subsequent analyses, and the environmental and health impacts.[16]

Lessons Learned from Chernobyl

Reactors in the United States and the West in general have different plant designs, broader shutdown margins, robust containment structures, and operational controls to protect them against the combination of lapses that led to the accident at Chernobyl. Thus, from a Western perspective, the Chernobyl accident could be dismissed as "different technology" run by a completely "different organization." However, the accident demonstrated some lessons that are relevant also for different and safer reactor designs.[17]

1. Three crucial elements are: containment; effective severe accident management strategies; and, perhaps most important, an inherent and/or passive safety function that can respond with no operator action for a set period of time.
2. Chernobyl demonstrated the importance of operator training, already brought home by TMI, and the complementary need for accurate and timely information about the complete reactor state available to the operators. As a result, a "global INPO" was agreed upon and WANO was established.
3. Precursor incidents that are not in themselves damaging but point to conditions that could lead to a much worse accident must be acted upon. In the case of Chernobyl, the International Nuclear Safety Advisory Group (INSAG-7 noted that "observations made at the Ignalina [Lithuania] plant in 1983, when the possibility of positive reactivity insertion on shutdown became evident, and the event at the Leningrad nuclear power plant in 1975 pointed to the existence of design problems. . . . [T]his important information was not adequately reviewed and, where it was disseminated to designers, operators and regulators, its significance was not fully understood and it was essentially ignored."[18]
4. Another important impact of Chernobyl was the realization that reactor accidents can have a regional impact on environment and health and a global impact on plans for future additions to nuclear power.

The above lessons learned were only partially acted upon, for a variety of reasons. Thus, with respect to lesson 1, while all new reactors have effective secondary containment features, better passive safety features—found in so-called Gen III+ plants—have only very partially been implemented. Their more complicated licensing and the associated higher financial risk have slowed their introduction, with most reactor vendors continuing to offer evolutionary reactor designs with active safety systems. In the case of lesson 2, the provision for better operator training, WANO's lack of teeth has led to an organization mainly devoted to sharing information, a necessary but insufficient feature. Additionally no effective carrot, e.g., through financial incentives, has been established. In contrast, INPO ratings are used by the financial community to assess U.S. utility stocks and by insurance companies to determine premiums. This is discussed further in the next section. With respect to lesson 3, precursor incidents, as demonstrated below in the discussion of Fukushima and Le Blayais, are still being overlooked in some cases. Lesson 4 has been generally internalized by the established nuclear power users but it remains to be seen whether it will also be internalized by the new users.

Fukushima-Daiichi

The March 2011 large-scale industrial accident at the Fukushima-Daiichi Nuclear Power Plant was the culmination of three interrelated factors: external natural hazard assessment and site preparation, the utility's approach to risk management, and the fundamental reactor design. The Fukushima-Daiichi Nuclear Power Plant was first commissioned in 1971 and houses six boiling water reactors (BWRs) ranging in size by age.[19] The reactor accident was initiated by a magnitude 9 earthquake on March 11, 2011, followed by an even more damaging tsunami. However, it was the inability to remove the decay heat in the reactor core that led to core meltdown and radioactive release from three units. The plant first experienced a station blackout (i.e., loss of all offsite and onsite power) due to flooding of backup critical emergency electrical generation equipment. Following failure of backup water injection equipment, delays in initiating injection of seawater into the reactors using portable pumping equipment led to the fuel overheating. Subsequently, the generation of hydrogen through steam oxidation of the fuel cladding led to chemical explosions causing significant structural damage.

While the public health impact has been low to date, the economic and nearby environmental consequences are severe. There is no doubt that land restoration will take over a decade and perhaps much longer. It will be several months to years before we will be in a position to learn most of the lessons from this tragedy. Nevertheless, several conclusions about preventive design, mitigation actions, and emergency response have been drawn by international organizations. It is unclear whether licensees will become more willing to make million-dollar investments to save billion-dollar assets and tens of billions of dollars of damages.

Lessons Learned from Fukushima-Daiichi

Unlike the other two reactor accidents, we have only limited hindsight as we look at drawing lessons from Fukushima-Daiichi. Three months following the accident, the Japanese government issued a report to the IAEA Ministerial Conference on Nuclear Safety.[20] In this report, the Japanese government identified twenty-eight lessons learned from the accident thus far. The NRC's Near Term Task Force[21] released its findings after a ninety-day review.[22] The NRC task force made twelve recommendations after reviewing the Fukushima-Daiichi accident. It attempted to structure its review activities to reflect insights from previous agency lessons-learned efforts. For example, some post-TMI recommendations considered a number of actions that were proposed for general safety enhancement as opposed to specific safety vulnerabilities revealed by the accident. The NRC Backfit Rule[23] may play an important role in determining which recommendations ultimately are implemented in the United States.

As with TMI and Chernobyl, we will not review each of the recommendations made by these organizations but rather identify dominant themes. It is important to note that many of the recommendations made by the Japanese government include major organizational changes that could be considered country-specific. In particular, recommendations on regulatory independence and emergency preparedness have already been implemented in many other countries but certainly not all. Additionally, many of the recommendations discussed below have not yet reached final approval.

The recommendations made by the NRC task force were divided into general regulatory philosophy concerns, ensuring protection, enhancing mitigation, strengthening emergency preparedness, and, finally, improving the efficiency of the regulatory oversight process of the fleet. Key themes from this task force set of recommendations and the Japanese government report are described below.

1. Each report acknowledged the need to rely on a defense-in-depth philosophy whereby resources must be allocated to measures that improve system protection, mitigation, and emergency response.
2. The Fukushima-Daiichi accident made global licensees and regulators re-evaluate whether their facilities have adequate protection from natural phenomena within the design basis. Additionally, a re-definition of the design basis and the way in which external hazards are treated was a constant theme.
3. The accident also made global licensees and regulators reconsider how much protection a plant should have from phenomena beyond the design basis.
4. A station blackout where all onsite and offsite AC (alternating current) power is unavailable has long been known as a highly vulnerable plant operational mode. Regulators make licensees demonstrate that the plant can meet an "acceptable" specified duration of time known as "coping time." A plant's coping time varies, depending on the redundancy and reliability of both onsite AC backup and offsite power options. The process is currently performance-based and risk-informed in the United States. However, Fukushima-Daiichi illustrated the importance of adequately defining an acceptable coping time.
5. There were some positive lessons from Fukushima-Daiichi. The effective performance of fission product scrubbing in the wet well that greatly reduced aerosol fission product release was impressive. We know from data collected by authorities (for example, measurements of the uptake of iodine-131 in children living near Fukushima) that the overall public health impact from the nuclear accident will be substantially smaller than the impact from the earthquake and tsunami.[24]
6. Both reports recognized the challenges posed by multi-unit accidents as opposed to a single-unit accident like TMI. NRC safety

inspections of the domestic fleet revealed that some sites were underprepared for a multi-unit reactor accident.[25]

7. Dominant themes in the Japanese government report were: assignment of responsibilities; chain of command from the highest relevant authority to the operators on the ground; and communications, important in every situation. The Japanese government recognized that there were critical communication failures on multiple levels[26] including the communication between local and central organizations, the communication to the public, and the communication to international organizations and the rest of the world. According to the report, "The Japanese Government could not appropriately respond to the assistance offered by countries around the world because no specific structure existed within the Government to link such assistance offered by other countries to the domestic needs."[27]

The Fukushima accident continues to have a major global impact. The three lines of rationalization noted at the beginning of this discussion cannot be used here: reactors of the same design as the ones at Fukushima can be found around the world; operations in Japan are not qualitatively different from those elsewhere; and while the phenomenology involved in the reactor is understood, that involved in such external events as earthquakes and tsunamis is not known precisely enough to permit prediction. The impact on Japan—where the whole nuclear industry, which provides over 30 percent of electrical power, has come under question—seems to be the most severe. But the impact is not limited to Japan. How many of the lessons will be learned cannot now be assessed.

Reactor Incidents

In this section, we review two critical reactor incidents that provide interesting insights into industry learning. In neither case were there health or environmental consequences; but the responsible licensee and regulator were caught significantly off guard in both. In the case of Davis-Besse, the trustworthiness of the industry was brought into question.

Criminal charges were filed and two employees and a former contractor were indicted for hiding key evidence from the regulator.

Davis-Besse Reactor Vessel Head Degradation

The Davis-Besse Nuclear Power Station in Oak Harbor, Ohio, closed down on February 16, 2002, for routine refueling and maintenance. During inspections, a refueling outage team discovered serious material flaws in the Control Rod Drive Mechanism located in the upper Reactor Pressure Vessel (RPV). Davis-Besse is a single 889 MWe pressurized water reactor, a Babcock & Wilcox design that was first commissioned in 1978. The penetrations were made of Alloy 600, which is a common material used to fabricate various parts and components in nuclear power plants and which has historically been susceptible to primary water stress corrosion cracking (PWSCC).

The extent of the pressure vessel corrosion, known as wastage area, was found to be approximately the size of a football. In some regions, instead of the original six-inch-thick reactor head, only the remaining three-eighths-inch stainless steel cladding inner liner made up the primary system pressure boundary. If the liner had failed, the plant would have undergone a loss of coolant accident (LOCA) and would have required activation of the emergency core cooling system to bring the reactor to acceptable standby conditions. With the degradation occurring so closely to the control rod penetrations, there was also considerable concern about the reactivity shutdown capability of the plant following a breach in the vessel.

In 2006, two former employees and a contractor were indicted after being criminally prosecuted for a series of safety violations and intentional cover-ups. It is important to note that while Davis-Besse was affected most by this incident, the failure of inaction represented a breakdown of the licensee (FirstEnergy), the NRC, and the INPO. The entire United States PWR fleet was consequently affected strongly. The degradation of the Davis-Besse upper reactor pressure vessel head was ultimately rated as a 3 on the INES scale and is classified as a serious incident.

Lessons Learned from Davis-Besse

Immediately following Davis-Besse, the NRC established a Davis-Besse Lessons Learned Task Force in order to better understand how such a failure in regulation could occur. On September 30, 2002, the task force reported[28] its findings to a senior management review team. The report included fifty-one recommendations for the NRC to take action. All but two of these recommendations were ultimately approved by the commission. These recommendations were placed into four categories: (1) assessment of stress corrosion cracking; (2) assessment of operating experience, integration of operating experience into training, and review of program effectiveness; (3) evaluation of inspection, assessment, and project management guidance; and (4) assessment of barrier integrity requirements.

An interesting point was raised in the Lessons Learned Task Force when it was learned that certain relevant operating experiences from other countries, involving similar RPV penetration nozzles, were not widely known within the NRC and the U.S. nuclear industry. In some cases, this experience was erroneously determined to not even be applicable to U.S. PWR plants.[29]

The fundamental issue—better understanding the governing phenomenology driving nickel-based alloy nozzle stress corrosion cracking—has constantly plagued the industry. In the spring of 2003, just a year after the Davis-Besse incident, apparent boron deposits were detected at the lower RPV head of South Texas Project unit 1 (STP-1) near two bottom-mounted instruments (BMIs). While this degradation was unexpected, the advancements made in visual examination of Alloy 600 components following Davis-Besse contributed greatly to locating these flaws.

The important takeaway from Davis-Besse was that the key regulatory and operational stakeholders involved failed at organizational, management, and technical levels. There had been a number of indicators of corrosion but they were not acted upon, probably owing to the priority being given to continued production. The deceit and resultant cover-up efforts helped corrode public confidence in the industry, representing a low point in U.S. commercial reactor operational history. The Davis-Besse event was driven by a priority of production over safety. Unexpected degradations such as what happened at STP will continue to occur. It is the licensee's resultant actions that matter.

Le Blayais Flooding

The Blayais nuclear plant is a complex of four 900-MWe pressurized water reactors built in 1981–83 alongside the Gironde marine estuary, the outlet for the river Garonne to the Atlantic Ocean in southwestern France. Major floods have been recorded in the area for centuries. EdF, the owner-operator, had put in place sea walls ranging in height from 4.75 to 5.2 meters[30] and taken other precautions prior to the incident that took place in December 1999. In the month before the incident, the plant annual safety report announced a planned increase in the height of the sea walls to 5.7 meters scheduled for the following year, though EdF delayed construction.

On the night of December 27, 1999, a combination of high tide, high waves driven by winds up to 200 km/hour (160 mph), and intense rain resulted in flooding and the loss of most power supplies, shutting the plant down. Diesel backup generators started up, maintaining power to plants 2 and 4 until some supply was restored. In unit 1, one set of the two pairs of pumps in the Essential Service Water System failed due to flooding; had both sets failed, the safety of the plant would have been endangered. In both units 1 and 2, flooding put part of the Emergency Core Cooling System out of commission.[31]

Owing to continued operation of some pumps and generators, cooling continued and the safety of the plant was not impaired. It was a close call, however, scored as Level 2 in the INES.[32] The incident had an impact on both political authorities, especially locally, and EdF.

Lessons Learned from Le Blayais Flooding

EdF and various advisory committees conducted a review that lasted seven years and focused mainly on the effects of combinations of adverse events, such as had led to the Blayais flood. As a result of the reviews,[33] protection against floods was upgraded in most French plants considered to be at risk, including higher dikes and seawalls, better sealed doors and closures, and a stricter protocol for protective action upon warning. A continuing assessment of the possible effect of climate change was also provided for. Total cost, estimated at 110 million euros, was small in light of annual EdF revenues and profits.

While EdF apparently learned some of the important lessons from the incident and has a continuing review process, what other countries learned is a question mark. Most obviously, while the Tokyo Electric Power Company (TEPCO) faced a far worse environment at Fukushima in the wake of the Tohoku earthquake than the Blayais environment, some of the lessons of Le Blayais were relevant. Most notable among these were improving the protection of backup power supplies (about which TEPCO had been warned by the Japanese regulatory authority) and establishing and rehearsing a clear protocol for dealing with flooding. In the United States, the Fort Calhoun Nuclear Generating Station on the Missouri River (about twenty miles north of Omaha, Nebraska) was surrounded by water up to a level of nearly 1,007 feet above sea level in June 2011. The protective berms and walls were 1,009 feet above sea level; the NRC had mandated an increase to 1,014 feet, which had been contested for a time by the operator, Omaha Public Power District. Similar levels had been reached in 1952; levels just short of 1,000 feet had been reached several times since.[34]

Flooding is only one potential external initiator for accidents, but it is an important one given that nuclear plants are frequently located near large bodies of water. Flooding risks are of particular concern as they are susceptible to a "cliff-edge" effect, in that the safety consequences of a flooding event can increase greatly with a modest increase in the flooding level.[35] These incidents and other lesser ones show two common features: the maximum design basis flood in some countries is uncomfortably close to floods that recur on a regular basis and climate change is likely to change the recurrence pattern of high waters and high winds (though not, obviously, of tsunamis). This preliminary look raises the question of whether flood protection should again be reviewed and should be a major part of protecting any new installation.

Reactor Anomalies

In this final section we will look at a reactor event that would be classified as an anomaly or an abnormal occurrence rather than an accident or incident. As mentioned earlier, anomalies and reliability indicators are very important as they can often serve as precursors for much larger incidents. The NRC recognizes this and requires itself to report to Congress annually about each abnormal occurrence for the fiscal year. The NRC defines an abnormal occurrence as an unscheduled incident or event that the regulator determines to be significant from the standpoint of public health or safety.

Northeast Blackout

On August 14, 2003, the largest blackout in the history of North America left 50 million people across southeastern Canada and the northeastern United States without power. Approximately half a year later, after a three-month investigation, a United States-Canada task force determined that a combination of human error and equipment failures were the root causes of the blackout.

Nine nuclear power plants tripped in the United States: eight plants lost offsite power and one plant was in an outage. The maximum amount of time until power was available to the switchyard for any plant was six and a half hours. While all onsite emergency diesel generators performed as designed, this event was significant due to the number of plants affected by the outage and the unexpected amount of time without offsite power.

Lessons Learned from Northeast Blackout Incident

The NRC immediately took action following the blackout incident by issuing a regulatory issue summary reminding licensees that they are required to comply with their technical specifications relative to inoperability of offsite power. The NRC also issued a generic letter[36] titled "Grid Reliability and the Impact on Plant Risk and the Operability of Offsite Power." It required licensees to submit information in four areas: (1) use of protocols between the plant and the transmission system operator (TSO) or independent system operator (ISO) and the use of transmission load flow analysis tools to assist plants in monitoring grid conditions to determine the operability of offsite power systems, (2) use of plant protocols and analysis tools by TSOs to assist plants in monitoring grid conditions for consideration in maintenance risk assessments, (3) offsite power restoration procedures, and (4) losses of offsite power caused by grid failures at a frequency equal to or greater than once in twenty site-years per regulation.

The NRC and the Federal Energy Regulatory Commission (FERC) have annually held joint meetings since the blackout incident to ensure adequate progress has been made in raising loss of offsite power capabilities of the domestic fleet.[37] Licensees and the NRC are routinely in communication with TSOs and ISOs in order to anticipate potential issues before they happen. The NRC also developed improved operator examination and training programs that gave operators practice in communicating with grid operators. The relationships among FERC, the North American Electric Reliability Corporation (NERC), the NRC, and domestic licensees appears to be proactive and will be further examined as the NRC recommendations from Fukushima are finally implemented.

Some Key Observations

What can be taken away from the foregoing retrospective survey of the more serious nuclear accidents and near-accidents and from the lessons learned—and not learned—from those events? A few observations emerge.

On the record of the past fifty years, nuclear power has an edge over other forms of providing energy both in limiting day-to-day adverse health and environment effects, including greenhouse gas (ghg) emissions, and in the frequency and toll of major accidents. Table 6.1 makes this point clear.

The low morbidity is due to several factors, but two stand out:

- Most casualties and other health and environmental effects stem from the extractive and transportation industries. Since the same

Table 6.1 Main Sources of Electricity in the World and Their Morbidity and Greenhouse Gas Emissions Per Unit Electricity Produced[38]

Source (% of World Use 2007)	Deaths Per Terawatt-hours	Tons ghg Per GWh (Life)
Coal (42%)	161 (U.S. 15)	800–1400
Gas (21%)	4	300–500
Hydro (16%)	0.1 (Europe)	Small–100
Wind (<1%)	0.15	Small–50
Nuclear (14%)	0.04	Small–50

amount of electric power can be obtained from about 100 tons of uranium ore as from 3–4 million tons of coal or similarly large quantities of gas or oil, these effects are inherently less for nuclear power than for the main hydrocarbon sources of electricity.

- The nuclear power industry has from the start been aware of the need for a strong and continued emphasis on the safety culture, although in the early years that culture was not sufficiently informed by experience. Even today, the need to meet performance goals has interfered—and too often continues to interfere—with best worker safety measures.

The low nuclear plant emissions of all kinds stem from the fact that no combustion is involved in nuclear electricity generation. Emissions are only generated during construction, installation, mining, refining, enrichment, transportation, and decommissioning. In addition, a nuclear reactor only needs on the order of 250 tons of raw ore per year while a coal reactor, e.g., needs between three and four million tons. This lowers emissions ascribed to nuclear plants due to mining, transporting, and processing ore. The actual amount of ghg generated depends on how the energy for the steps outlined is obtained as well as on the techniques used to make the needed concrete.

Despite all these advantages, accidents will always be possible. The process of learning to improve safety from experiencing accidents, close calls, and routine problems should be viewed as a continuing investment in assuring both the political and financial future of the nuclear industry. It has to be considered as part of the base levelized cost of power and it reaches every part of the process of providing nuclear power, from qualification of materials such as concrete and steel to operations. The knowledge to do this comes from experience. Most of it, fortunately, has been obtained from research and day-to-day learning without, for the most part, major accidents. It is, however, a never-ending process for nuclear power and for all complex engineered systems that have the potential for causing major disasters.

All three of the major nuclear power accidents (TMI, Chernobyl, and Fukushima) as well as several of the less well-known close calls had precursors in previous incidents, although often not in the same country. The lessons-learned reviews that have followed most of these events usually made specific useful points. Some of those were implemented—the lessons were learned—but often they were not. Not surprisingly, implementation steps that translated into more efficient operations, such as better, more standardized operating procedures, were carried out more often than steps that required immediate expenditures, such as better defenses against flooding, to avoid uncertain disaster. Further analysis may reveal less obvious correlations.

A regulating agency with appropriate power and strong technical expertise—well-staffed, well-funded, and independent of its licensees—is a necessary, though not sufficient, requirement for safety and in particular for the formulation and implementation of lessons learned. Regulator capture by licensees through either political or administrative processes has been a problem in several countries. Recent moves, e.g., in India, to remove the regulating agency from the administrative structure of the operating and promoting agency is a move toward greater safety. Beyond an effective regulator, however, a culture of safety must be adopted by all operating entities. For this, the tangible benefits of a safety culture must become clear to them.

In the United States, and in some other countries, public fear of radioactivity and the ensuing interventions of often well-informed organizations have been a spur to learning from experience. But in countries where the responsible nuclear organizations, governmental and private, were insulated from criticism, learning has been slower. Learning from experience is never an easy process, especially when the learning process takes place in a very public, very critical arena. Nevertheless, transparency has helped that process. Transparency helps learning in all three groups: the owners-operators, the government regulators, and some of the intervening organizations. Transparency must be conditional, however: the early critical give-and-take that leads to improvements in design, materials, and operation will not be done frankly and effectively if it is not done in private.

INPO, funded and supported by the U.S. nuclear power industry, is an example of a well-balanced combination of transparency and privacy that provides an ongoing process for learning lessons in the operations area. Operator ratings at the various plants remain private but results in terms of operating procedures and consequences are public. INPO was started in the United States as a result of the TMI accident. The accident, as discussed above, resulted from design deficiencies, lack of understanding of some fundamental phenomena, and errors in operating the reactor. Design features prevented any significant release of radiation to the public but the financial loss was so significant that the industry and its financial backers were moved to cooperate with regulators in establishing and maintaining much improved operator training, operations standards, and operations staffing. The resulting rating of the operators is kept confidential within the industry so that criticism can be uninhibited and action can be taken in a timely way without fear of misinterpretation. On the other hand, actual performance results, including all incidents, are public. Expert management in the owner-operator sector has been essential to establish and maintain quality of operations. In addition, dealing with reactors during abnormal conditions requires well-thought-out procedures, clearly established lines of authority, and onsite personnel competent and authorized to make tough decisions.

Because so much of the cost of nuclear power is incurred before the first kilowatt-hour is generated, the financial backers, including private and government insurers and guarantors, have in theory considerable leverage over the industry, as does any entity that can delay construction and operations such as regulators and interveners. That leverage can be obvious, as when the European Reconstruction Bank refused to put money into older Chernobyl-type reactors but insisted on safer Western-style models; but it must work with a regulating and monitoring institution to preserve the investments. Recognition of the need for such a strong, competent, and independent regulator has not come easily to most countries and is not always and everywhere accepted in practice to this day. In particular, independence (plus resources sufficient to maintain competence) runs into continuing tensions from operators (and, in democracies, their representatives in government), who need to make a profit or at least stay within budget while maintaining market share and also from government budgeters who have to work with limited overall resources.

Looking to the Future

Examining the above conclusions and looking to the future of nuclear power worldwide, we come to another set of observations.

1. *Modern reactors are of safer designs and can be operated more safely than the ones that have caused major accidents. But it is not clear how many of the safest designs will be built.* Most reactors being built today are of the "Gen II +" design and are significantly safer than the RBMK design involved in Chernobyl and the Mark 1 BWR design involved in Fukushima in that both the reactor vessel and the spent fuel are under two layers of containment. Even safer designs, such as the Gen III and Gen III+, feature more passive cooling systems, which can keep all fuel cool for days without electricity or high-pressure water injection, along with other improvements. It is unclear at this writing what the future reactor mix will be. An interesting question is whether new reactor users will buy modern designs while the existing users will mostly extend the lifetimes of their existing designs.

2. *The Fukushima accident was caused by a "one-in-a-thousand-years event" plus some admittedly faulty design and siting. The precursor incident at the Blayais nuclear plant in France had also been viewed*

as a "one-in-a-thousand-years event." Nevertheless, looked at on a worldwide basis and considering the proposed lifetimes of modern reactors, there is a clear statistical basis for taking such events into account and spending some money to prevent or alleviate their consequences. Reactor lifetimes today are of the order of sixty years, which is 6 percent of the "thousand years" postulated for the recurrence time of the Tohoku tsunami. The likely cost of the Fukushima accident is estimated at $30 billion to $100 billion. Taking into account these figures, the number of sites subject to tsunamis, and the worldwide impact of a Fukushima-like accident, much could justifiably be spent to better alleviate the consequences of very rare events. While tsunamis are not the only possible external source of disaster and while prioritization in allocating limited resources is always needed, a new look at rare but potentially catastrophic events and the precautions that could be useful in dealing with them could lead to new measures that would help deal with other categories of rare events and would be economically justifiable.

3. *Mechanisms to facilitate and, where needed, enforce mutual learning among countries are not as effective as they are within countries and may not be adequate to prevent avoidable disasters.* The present mechanisms are unsystematic and do not have enforcement or incentive features. They include the efforts of vendors to build safer reactors, the general availability of lessons learned from particular accidents and near-accidents, and awareness of the worldwide cost of a nuclear accident anywhere. At the institutional level, the two active organizations are the IAEA and WANO. The IAEA produces reports and submits protocols for adoption by its nation-members. It has major responsibilities in other areas, e.g., safeguards against military use of civilian facilities, and does not have enough personnel, budget, or authority to set and enforce safety standards, should those be agreed upon. WANO's focus is on reactor operation, an essential—but not the only—ingredient of safety. Its main activity there is information-sharing. INPO,

its U.S. counterpart, discussed above, is quite effective. But it is a confidential and cooperative U.S. industry effort that seems difficult to duplicate on a worldwide basis, at least without major changes.

4. *Improved cooperation will rest most securely on lasting shared economic interest among vendors, owners-operators, government regulators, and the public. At the same time, the international nuclear power and nuclear fuel cycle markets will become, if anything, more competitive than they have been.* No solution to this problem is in sight. Elements of a solution that might be considered include the following factors.

    - Some form of import-export agreement, such as the Nuclear Suppliers Group (NSG) now uses to monitor weapons-sensitive materials and components, might be effective. Those efforts rest on an agreement at the state level; the same would be true of a safety-oriented agreement. If there were such an agreement among states, one could envisage that any vendor wishing to export reactors or any other potentially dangerous nuclear facility would need a license certifying that the design meets modern safety standards. There are only a few international reactor vendors, so that implementation might be feasible.
    - Reactor design is not the only ingredient of safety. Siting, construction practices, and operations also enter in essential ways, as do accident management, regulatory review, and lessons-learned feedback. Agreement at the state level strengthening cooperation among regulatory authorities—perhaps even setting standards for independence of those authorities—would be a step to meet problems there. There is no clear consensus on what structure best assures such independence—or, rather, effectiveness—at managing an inherently interdependent process that involves many stakeholders. A conversation on the subject that would take into account national precedents and institutions is needed before any attempt is made at discussing standards.

- Finally, investors and insurance companies have strong incentives to avoid serious accidents. Insurance company liability is generally limited, leaving investors and taxpayers to take losses. In most countries, investment comes in part from government, in part from bond sales. Investment represents a potential source of leverage to avoid accidents but to date it has not been harnessed toward effective action. This occurs in part because of a lack of knowledge and in part because nuclear-related investments may be only a small part of the portfolio.

Notes

1. Richard A. Meserve, "The Global Nuclear Safety Regime," *Daedalus,* vol. 138, no. 4 (Fall 2009), pp. 102ff. Richard Meserve is a former chairman of the U.S. Nuclear Regulatory Agency (NRC).

2. Licensees are the entities licensed to construct, operate, and otherwise deal with nuclear installations. They are mainly electric utilities but also include non-profit research organizations and others.

3. WANO is an international advisory body. EPRI is an independent research organization of the U.S. utilities.

4. William Keller and Mohammad Modarres, "A Historical Overview of Probabilistic Risk Assessment Development and its Use in the Nuclear Power Industry: A Tribute to the Late Professor Norman Carl Rasmussen," *Reliability Engineering & System Safety,* vol. 89, no. 3 (2005), pp. 271–285.

5. Some of this was facilitated by EPRI and involved cooperation between utilities and the nuclear industry, leading to new advanced reactor designs that took advantage of lessons learned from prior incidents.

6. The NRC is structured to function as an independent agency, in which commissioners can only be removed for just cause. In most executive branch agencies, the administrators serve at the will of the president of the United States.

7. One NRC spokesperson put this point a different way: "A really good careful driver can probably drive a poorly designed car with no bumpers, but a poor driver can easily wreck a well-designed car." See Joseph V. Rees, *Hostages of Each Other: The Transformation of Nuclear Safety Since Three Mile Island* (Chicago: University of Chicago Press, 1994).

8. The "Rogovin Report" disputes the role of operator error as a major contributor to the TMI accident. Instead, it cites inadequate training, poor operator procedures, a lack of diagnostic skill on the part of the entire site management group, misleading instrumentation, plant deficiencies, and poor control room design. Whatever the cause, some operator actions clearly contributed to the accident.

9. Douglas M. Chapin, et al. "Nuclear Power Plants and their Fuel as Terrorist Targets," *Science,* vol. 297, no. 5589 (September 20, 2002), pp. 1997–1999. Also, Mitchell Rogovin and George T. Frampton Jr., "Three Mile Island: A Report to the Commissioners and to the Public," Nuclear Regulatory Commission, Special Inquiry Group (Washington, D.C.: U.S. Government Printing Office, 1980).

10. U.S. Nuclear Regulatory Commission, "TMI-2 Lessons Learned Task Force Final Report" (NUREG-0585), Washington, D.C., 1979.

11. John G. Kemeny, *Report of the President's Commission on the Accident at Three Mile Island* (New York: Pergamon Press, 1979).

12. Joseph V. Rees, *Hostages of Each Other: The Transformation of Nuclear Safety Since Three Mile Island* (Chicago: University of Chicago Press, 1994).

13. Contrast this mindset with Admiral Rickover, the "Father of the Nuclear Navy," who famously once said, "My program is unique in the military service in this respect: You know the expression 'from the womb to the tomb'; my organization is responsible for initiating the idea for a project; for doing the research, and the development; designing and building the equipment that goes into the ships; for the operations of the ship; for the selection of the officers and men who man the ship; for their education and training. In short, I am responsible for the ship throughout its life—from the very beginning to the very end." See "Hearings on Military Posture and H.R. 12564," Department of Defense authorization for fiscal year 1975, 93rd Cong. 2nd sess. (Washington, D.C.: U.S. Government Printing Office, 1974), p. 1392.

14. Rees quotes former University of California-Berkeley Professor Tom Pigford: "The massive effort to comply with the vast body of [NRC] requirements and to demonstrate compliance therewith . . . foster[ed] . . . [the] complacent feelings that all of the work in meeting regulations must somehow insure safety."

15. NRC, "TMI-2 Lessons Learned Task Force Final Report."

16. For a generally accepted analysis of the sequence of events, the causative factors of the accident, and a summary of measures to improve the safety of RBMK reactors, see International Nuclear Safety Advisory Group, "The Chernobyl Accident: Updating of INSAG 1," Safety Series No. 75-INSAG-7 (1992), commonly referred to as INSAG 7, and references and annexes therein,

including to the earlier document, INSAG 1. The NRC backgrounder can be found at http://www.nrc.gov/reading-rm/doc-collections/fact-sheets/chernobyl-bg.html. For a description of the RBMK reactor and more details on safety fixes after the Chernobyl accident, see "RBMK Reactors" at http://www.world-nuclear.org/info/inf31.html. For a summary of environmental and health effects, see "The Chernobyl Accident: UNSCEAR's assessments of the radiation effects" at http://www.unscear.org/unscear/en/chernobyl.html and references therein, especially "Health effects due to radiation from the Chernobyl accident (2008)," an authoritative and detailed recent assessment. UNSCEAR is the United Nations Scientific Committee on the Effects of Atomic Radiation.

17. In what follows, we do not discuss the fixes specific to the RBMK. Those may be found in the references noted above, particularly "RBMK Reactors," which also has a list of currently operating RBMK reactors.

18. International Nuclear Safety Advisory Group, "The Chernobyl Accident: Updating of INSAG 1," Safety Series No. 75-INSAG-7 (Vienna: International Atomic Energy Agency, 1992).

19. The smallest and oldest unit 1 was 460 MWe while units 2–5 were 784 MWe. Unit 6 was the most recently built unit and was 1,100 MWe.

20. Nuclear Emergency Response Headquarters, Government of Japan, "Report of Japanese Government to the IAEA Ministerial Conference on Nuclear Safety—The Accident at TEPCO's Fukushima Nuclear Power Stations" (Vienna: IAEA, June 2011).

21. The NRC determined both short-term and long-term task forces should be established in a similar fashion to Japan.

22. Charles Miller, Amy Cubbage, Daniel Dorman, Jack Grobe, Gary Holahan, and Nathan Sanfilippo, "Recommendations for Enhancing Reactor Safety in the 21st Century: The Near-term Task Force Review of Insights from the Fukushima Dai-ichi Accident," Nuclear Regulatory Commission, July 12, 2011, http://pbadupws.nrc.gov/docs/ML1118/ML111861807.pdf.

23. The Backfit Rule was originally introduced into NRC rule-making in 1970. A later rule-making change required that a backfit "must result in cost-justified substantial increase in protection of public health and safety or common defense and security."

24. Per F. Peterson, Testimony to California State Senate Energy Committee Hearing on Nuclear Power Plant Safety, Panel on "Seismic and secondary seismic risks near nuclear power plants and spent fuel rod storage facilities in California," April 14, 2011, http://seuc.senate.ca.gov/sites/seuc.senate.ca.gov/files/04-14-11Peterson.pdf.

25. An example was inadequate mutual aid agreements. For example, Diablo Canyon Power Plant (DCPP) near San Luis Obispo, California, identified the fact that no memorandum of understanding was in place with the California National Guard for the contingency to supply diesel fuel to the site when the main road was unavailable. More examples can be found at the NRC investigation report: http://pbadupws.nrc.gov/docs/ML1113/ML11133A310.pdf.

26. As well as a poor delineation of responsibilities, at least in some observers' view.

27. Nuclear Emergency Response Headquarters, "The Accident at TEPCO's Fukushima Nuclear Power Stations."

28. For the final report, see U.S. Nuclear Regulatory Commission, "Davis-Besse Reactor Vessel Head Degradation," Lessons Learned Task Force Report (2002), http://www.nrc.gov/reactors/operating/ops-experience/vessel-head-degradation/lessons-learned/lltf-report.html.

29. U.S. Nuclear Regulatory Commission, "Davis-Besse Reactor Vessel Head Degradation," Lessons-Learned Task Force Report, 2002.

30. These are measured from NGF, a sea-level standard used in France.

31. Material taken from http://en.wikipedia.org/wiki/1999_Blayais_Nuclear_Power_Plant_flood; http://www.dissident-media.org/infonucleaire/page_blayais.html; and http://vert-estuaire-charentais.over-blog.com/article-27345494.html. Also see J. M. Mattéi, E. Vial, V. Rebour, H. Liemersdorf, and M. Türschmann, *Generic Results and Conclusions of Re-evaluating the Flooding in French and German Nuclear Power Plants,* Eurosafe Forum 2001, http://www.eurosafe-forum.org/files/semb1_7.pdf; A. Gorbatchev, Jean-Marie Mattéi, Vincent Rebour, and Eric Vial, "Report on flooding of Le Blayais power plant on 27 december 1999," Institute for Protection and Nuclear Safety, 2000; and Eric de Fraguier, Presentation on "Lessons Learned from 1999 Blayais Flood: Overview of EdF Flood Risk Management Plan," March 2010, http://www.nrc.gov/public-involve/conference-symposia/ric/slides/th35defraguierepv.pdf.

32. INES is a qualitative, somewhat subjective scale introduced by the IAEA in 1990 and intended to provide a logarithmic measure of the severity of a nuclear or radiological event.

33. Eric de Fraguier, Presentation on "Lessons Learned from 1999 Blayais Flood: Overview of EDF Flood Risk Management Plan," March 2010, http://www.nrc.gov/public-involve/conference-symposia/ric/slides/th35defraguierepv.pdf.

34. Peter Behr, "A Nuclear Plant's Flood Defenses Trigger a Yearlong Regulatory Confrontation," *New York Times*, June 24, 2011, http://www.nytimes

.com/cwire/2011/06/24/24climatewire-a-nuclear-plants-flood-defenses-trigger-a-ye-95418.html?pagewanted=all.

35. This observation was made in NRC's recently released near-term task force report on insights from the Fukushima-Daiichi accident.

36. http://www.ferc.gov/eventcalendar/Files/20060403161019-nrc-gl200602.pdf.

37. It should also be mentioned that the severe accident capabilities of a United States nuclear plant increased greatly following the 9/11 terrorist attacks. While these changes were targeted toward specific extreme external threats such as airplane attack and large fires, the capabilities of a plant's defense, mitigation efforts, and emergency response have greatly improved.

38. This table was generated from the following sources:

A. International Energy Agency, "Key World Energy Statistics 2009," p. 24.

B. Seth Godin, "Deaths per terawatt-hour," http://sethgodin.typepad.com/seths_blog/2011/03/the-triumph-of-coal-marketing.html.

C. Canadian Energy Research Institute, "Comparative Life Cycle Assessment (LCA) of Base Load Electricity Generation in Ontario," prepared for the Canadian Nuclear Association, 2008, http://www.cna.ca/english/pdf/studies/ceri/CERI-ComparativeLCA.pdf.

D. Benjamin K. Sovacool, *Valuing the greenhouse gas emissions from nuclear power: A critical survey*, Energy Governance Program, Centre on Asia and Globalisation, Lee Kuan Yew School of Public Policy, National University of Singapore.

E. Bert Metz, Ogunlade Davidson, Peter Bosch, Rutu Dave, and Leo Meyer, eds., *Climate Change 2007: Mitigation of Climate Change*, Intergovernmental Panel on Climate Change (Cambridge, UK: Cambridge University Press, 2007).

F. Alfred Voβ, *Energy and Sustainability: An Outlook*, Institute for Energy Economics and the Rational Use of Energy, Universität Stuttgart, International Materials Forum 2006, Bayreuth.

Nuclear Technology Development: Evolution or Gamble? 7

PER F. PETERSON AND REGIS A. MATZIE

Introduction

The move toward more advanced and innovative light water reactor (LWR) designs started in the mid-1980s as an initiative of the U.S. utility industry. Several visionary utility executives worked with the Electric Power Research Institute (EPRI) to document the requirements for future plants that would capture all the lessons learned over the previous thirty-five years of commercial operation, including the realization that in addition to public safety as an essential element in any future reactor design, reliable operation and investment protection have similar goals and importance. The Three Mile Island unit No. 2 (TMI2) reactor accident demonstrated that a multibillion-dollar asset could be lost even though very little radiation was released to the environment. What resulted out of this vision was the EPRI-sponsored advanced light water reactor (ALWR) utility requirements document program and follow-on U.S. Department of Energy first-of-a-kind (FOAK) engineering and design certification (DC) programs to advance several reactor concepts to the stage that they would be available for utility deployment. From these public/private collaborative programs came three designs that achieved DC: the General Electric advanced boiling water reactor (ABWR) in 1997, the Combustion Engineering System 80+ in 1997, and the Westinghouse AP600 in 1999. However, none of these designs were built in the United States, although ABWRs have been built in Japan and more recently in Taiwan.

In 2002, the Department of Energy (DOE) initiated its Nuclear Power 2010 program to cost-share with industry the effort to develop U.S. Nuclear Regulatory Commission (NRC) early site permit (ESP) licenses for three utility sites and to cooperate with industry to develop and license advanced nuclear plant technologies and demonstrate untested regulatory processes, including the combined construction and operating license (COL). As this program progressed, a key question faced by the increasing number of utilities considering the construction of new reactors involved what level and types of technical risk to take on in selecting reactor technologies. One strategy, commonly referred to as "evolutionary," emphasized the selection of designs that involved relatively small extrapolations from operating reactors, e.g., increasing the redundancy of "active" safety systems and implementing severe accident mitigating features based on the lessons learned from the TMI2 reactor accident. The other approach emphasized designs with much larger departures from the evolutionary path, in particular, the replacement of "active" safety systems with "passive" safety systems.[1]

Today, the first new nuclear plants that will start construction in the United States in over thirty years will be Westinghouse AP1000 reactors, which took the latter path of passive safety systems. Beyond passive safety, shown in Figures 7.1 and 7.2, the AP1000 also uses a much higher level of factory prefabrication and modular construction than any previous reactors, including the use of factory-prefabricated steel/concrete composite building structures that greatly reduce the site labor requirements and therefore provide greater certainty on both quality and schedule. While introducing these innovations required accepting and managing technical risk, construction of the first four AP1000 units in China remains on schedule and the construction of new AP1000 reactors in the United States began in 2012.

Whether nuclear energy will expand greatly, or instead show only modest increases or even contract, will depend greatly on whether significant future construction cost reductions can be achieved while continuing to meet evolving societal and regulatory requirements for safety, security, safeguards, and environmental protection. Other stimuli to nuclear expansion are the relative costs of fossil fuels and the premium that society places on emission-free power generation.

We argue that design standardization is essential to commercial competitiveness and offer a list of ten strategies that focus on the development of well-defined and highly standardized reactors as an essential element for commercial success and market penetration for nuclear energy. But we also argue that introducing technical innovation and taking on technical risk are key to commercial competitiveness. In fact, continuing introduction of technical innovation is likely essential to long-term commercial competitiveness. Innovation is likely to remain a major factor influencing the future economics and market penetration of nuclear energy.

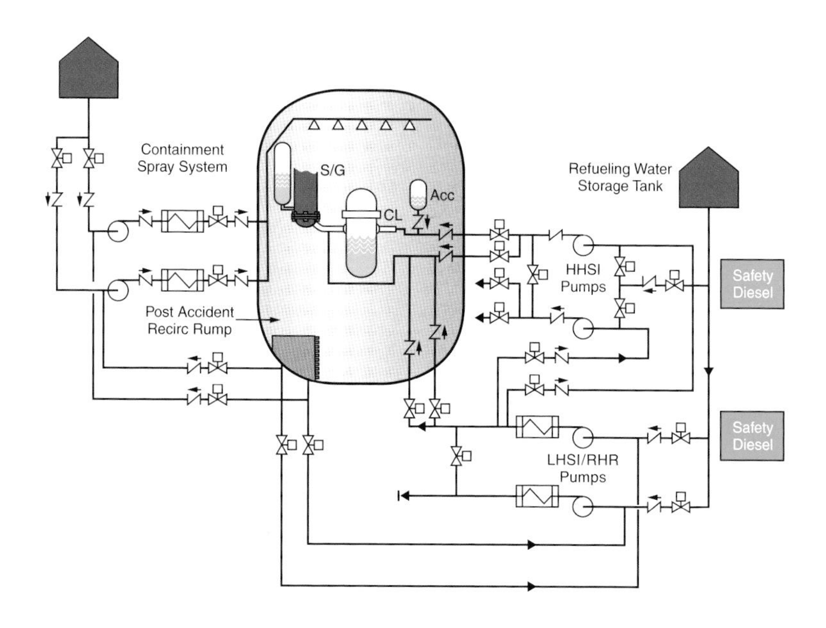

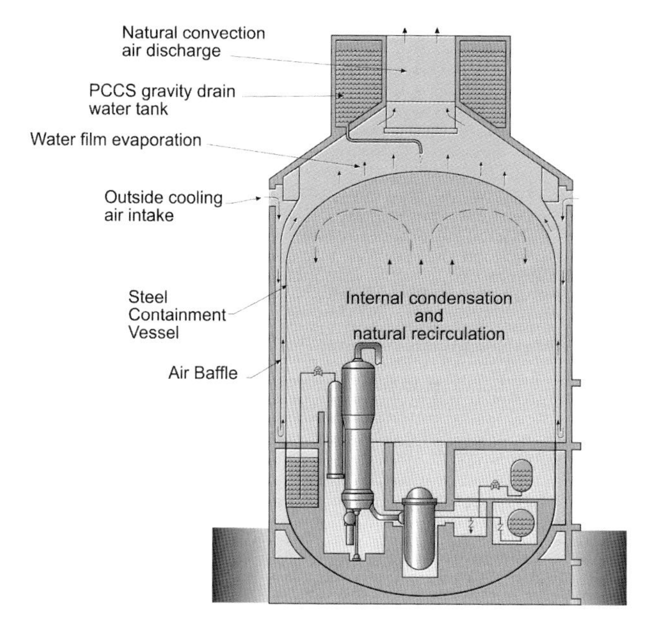

Figure 7.1 "Active" reactor safety systems (top) use a combination of heat exchangers and pumps (with associated electrical power supplies) to remove heat from a reactor core, while "passive" safety systems (bottom) accomplish this using natural forces such as gravity and evaporation without the need for electrical power input.
Source: Westinghouse Electric Company.

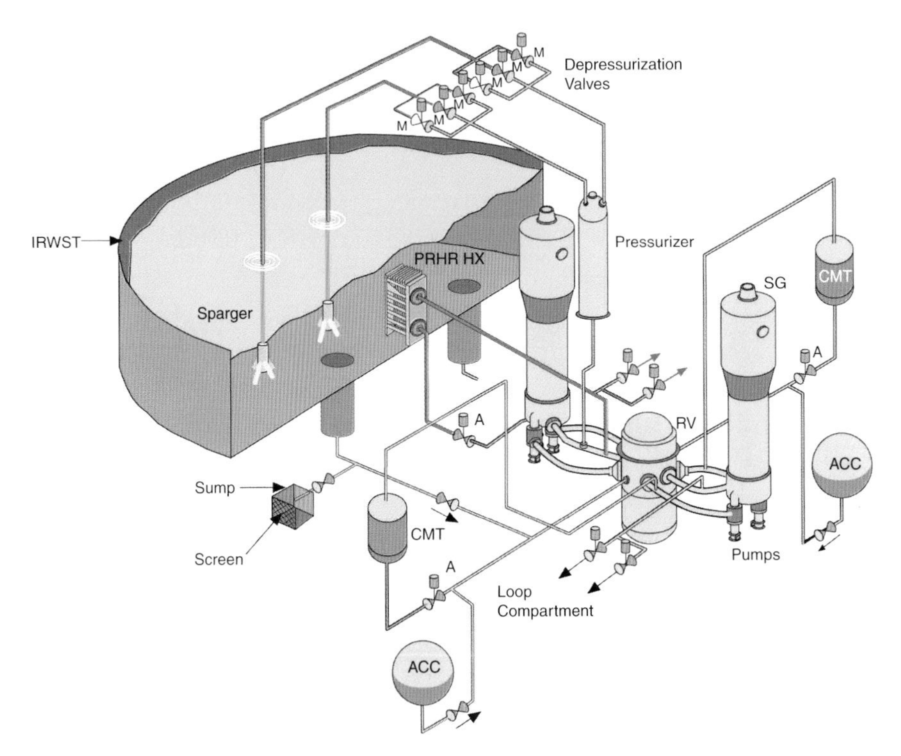

Figure 7.2 The passive safety systems of AP1000 consist of a collection of tanks, valves, heat exchangers, and their connecting piping and water inventories in containment, all available when called upon during a reactor accident to cool the core and keep it covered with water.
Source: Westinghouse Electric Company.

While the AP1000 provides an outstanding case study for the commercially successful introduction of innovation into a new reactor design, this paper is not intended to be a sales presentation for the AP1000 or to reveal Westinghouse's business strategy for developing future reactor technologies. Instead, this paper draws upon experience gained with the AP1000 and presents the major issues associated with achieving commercial success with new reactor technologies and the implications for both the challenges and the necessity to introduce further innovation in future reactor designs.

Essential Elements for Commercially Successful Reactor Deployment

The development of commercially successful nuclear reactors must involve a high degree of discipline in the design and licensing process, as well as in design completion and standardization. We argue that the ten main approaches to make future nuclear reactor development projects successful are as follows.

1. **The design must be *simplified* compared to plants operating today.** This has advantages in reducing detailed engineering design effort, construction schedule, overall plant cost, and operations and maintenance (O&M) requirements. This approach also makes it easier for operators to recognize upset conditions and take the proper action, thus contributing to improved safety. For example, the AP1000 uses "canned" primary pumps that have motor rotors inside the primary pressure boundary, eliminating the need for complex high-pressure shaft seals used in earlier plants. The elimination of normal seal leakage also improves the capability to detect coolant leaks at lower flow rates.
2. **Safety, security, and international safeguard functions should be harmonized by taking them into account early in the design process, as one key dimension of plant design simplification.** For example, the use of passive safety systems, as illustrated in Figures 7.1 and 7.2, can eliminate the need for frequent operator surveillance of equipment, allowing it to be hardened by placement inside the reactor containment structure to prevent easy access during security events.
3. **The commodities of the plant need to be minimized.** While raw material costs for nuclear plants are small, reducing commodities (as shown in Figure 7.3) helps shorten the construction schedule, reduces the number of craft workers at the site (who are difficult to manage in an efficient way), and thus reduces the cost.

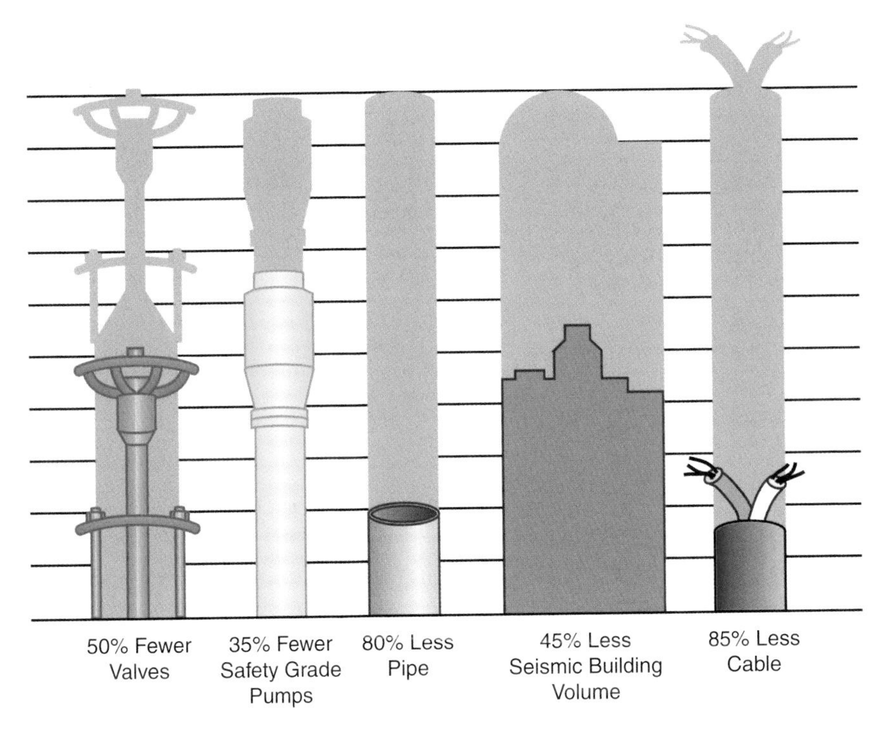

Figure 7.3 Simplification of the AP1000 design by the use of passive safety systems has resulted in a great reduction of commodities and therefore site labor and construction schedule.
Source: Westinghouse Electric Company.

4. **Reliability engineering should be included as a major element of reactor design.** This includes employing a systematic identification of failure modes and effects and designing components and systems to support on-line monitoring, in-service inspection, maintenance, and replacement. This approach complements the use of probabilistic risk assessment (PRA) techniques that help guide the evolving design toward improved safety by systematically eliminating weaknesses in the design that could lead to core damage.
5. **Plant design must strive for as much modularization as possible.** In particular, modularizing the structures is the place that gives great gains on the schedule. Modularization of equipment, piping,

etc., is conventional to many industries and now commonplace for new reactor construction in Asia. The use of structural modules, as in the AP1000, remains less common (cf. Figure 7.4).

6. **The design of the overall plant must be essentially complete before construction is started.** Obviously, there are site-specific aspects of the design for cooling water, intake structures, etc. But for the entire power block, the design can have a very high level of detail. This then allows for nth-of-a-kind (NOAK) costing on follow-on projects. This eliminates a lot of the FOAK surprises and rework that can occur if the design is not complete.
7. **The design must be pre-licensed by a very thorough and competent regulator before the project contract is signed.** Additionally, any changes must be at the cost and schedule of the owner. Design changes at the whim of the owner or the regulator must put the burden on the owner to pay for and argue against.

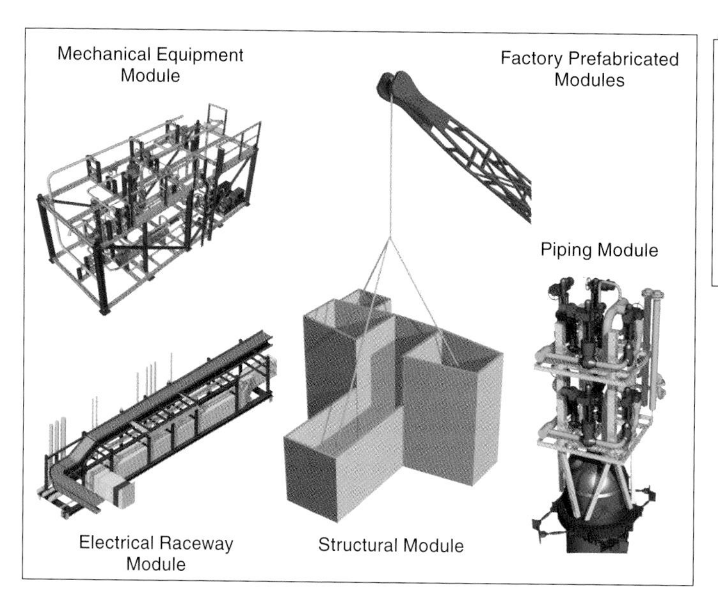

Module Type	Number
Structural	122
Piping	154
Mechanical Equipment	55
Electrical Equipment	11
TOTAL	342

Modules Designed Into AP1000

Figure 7.4 The different types of modules designed into AP1000 along with their numbers are shown in this figure. This approach has allowed the first AP1000 construction project to remain on schedule despite typical FOAK issues.
Source: Westinghouse Electric Company.

8. **Only standardized designs should be contracted.** There are obvious regional differences, e.g., sixty hertz versus fifty hertz, but these just establish a small set of standardized designs that can be implemented broadly. This argues against trying to modify a design for specific environmental conditions that are unique to a location. For example, to locate a plant on the Pacific Rim requires a very high seismic design, which then can become a new standard design for the entire region. However, a design should not be adapted for a specific site. If the site does not fit into the design envelope of the design, then it should be passed up.
9. **An execution team should be established that will be used repeatedly.** This includes the reactor vendor, architect-engineer (A/E), constructor, and major suppliers. Again, this is important for NOAK costing, for quality, for minimizing problems during construction, etc.
10. **The importance of writing a good contract with appropriate risk-sharing must be recognized.** The better the contract, the higher probability that the project will be successful—obviously subject to some, if not all, of items (1) through (9) above.

Post-Fukushima Reactor Safety

Fukushima was the first major reactor accident to be caused by an extreme natural disaster, rather than by some combination of equipment failures and human errors. Reactor designs have always acknowledged the risk from natural "external events" that include fire, flooding, seismic events, and environmental missiles, e.g., generated by tornados. These hazards vary from site to site, and the approach to assessing these hazards varies from nation to nation. More recently a few countries, particularly the United States, have also required that malicious external events, in particular crashes by large commercial aircraft, be considered. What the Fukushima accident has brought into question is the conservatism that should be used in establishing the parameters for these external events. What is the appropriate seismic intensity, what is the appropriate flood level, etc., against which the plant should be protected? The ability to cope with the consequences of these extreme external events will in the end be enhanced substantially as a result of the lessons learned from the Fukushima accident. In particular, the required coping time for a loss of all AC power—i.e., the time that the reactor can wait through the event with no core damage—will most likely be extended from what it is today (between four and eight hours for operating plants in the United States).

Another significant lesson learned is that procedures for managing and mitigating beyond-design basis accidents must be reviewed and practiced. These include severe accident mitigation guidelines (SAMGs), developed after the TMI accident, and extensive damage mitigation guidelines (EDMGs), developed in the United States after the 9/11 attack (but not adopted in Japan). EDMGs focus on procedures to restore key safety functions (water injection and power generation) using portable equipment while SAMGs focus on mitigating damage and controlling releases if fuel becomes damaged. Immediate action responsibilities, lines of authority, pre-positioned accident response equipment—all need to be reviewed and, if necessary, upgraded. Following a severe external event that causes extensive plant damage, information is a priority and without sufficient planning and preparation can be unavailable or insufficient. It is likely that some new instrumentation will be installed in operating plants that can be used by the accident management team to understand the accident progression and take the appropriate actions to, for example, measure spent fuel pool water level and temperature.

Emergency operating procedures (EOPs) are well understood and practiced by plant operators and management. However, EDMGs and SAMGs have had less attention and are generally not well understood or practiced. Because of the infrequency of use of the latter guidelines and the high importance of properly using them when necessary, more attention will need to be placed on familiarity and execution of these EDMGs and SAMGs.

Likewise, the Fukushima accident emphasizes the importance of prioritizing regulatory requirements by their risk significance. The fact that a reactor is operated in compliance with regulatory requirements does not necessarily mean that it is sufficiently safe, unless those regulatory requirements are grounded by risk assessment. In Japan, the regulatory authority requires that all reactors be shut down at least once every thirteen months to perform refueling, inspections, and maintenance. Modern practice in other countries places no deterministic regulatory limit on refueling intervals, so most plants refuel every eighteen to twenty-four months, resulting in increased plant availability and improved economics. The fact that the regulatory authority in Japan placed an uncommonly stringent and expensive limit on refueling intervals, compared to international practice, arguably contributed to complacency about nuclear plant safety.

It should be noted that modern ALWRs would have fared better during the Fukushima accident than the 1970s-era Generation II designs at that site. The reason for this is the greater level of redundancy of the safety systems and the greater protection of these safety systems against external events, in particular flooding. The Generation III+ designs like AP1000, in fact, would have ridden through the seismic event and accompanying tsunami because of their passive safety systems that do not rely on AC power. They have all necessary cooling water available within the plant, they have greater battery capacity, and they rely on natural processes, such as evaporative cooling, gravity, and conduction heat transfer, to provide the safety functions. These capabilities are designed to function for a minimum of seventy-two hours without operator action. Then only simple actions are required such as the alignment of accessible valves to bring other sources of available water on-line. In addition, all Generation III and III+ ALWR designs already have a suite of severe accident mitigation systems to combat hydrogen generation, molten core-concrete interaction, high pressure corium ejection, etc. Thus, many of the consequences that were of extreme worry at Fukushima would have been addressed by available systems in these new plant designs.

One other apparent lesson learned from the Fukushima accident is that spent fuel might need to be managed in a more conservative way. Fukushima had spent fuel stored in pools located at high elevations inside each reactor secondary containment, in a below-grade central pool, and in dry casks. The below-grade central pool and the dry cask storage both performed very well. The elevated spent fuel pools inside the reactor buildings did not have any wide-range level instrumentation, and because the pools became physically inaccessible, very large uncertainty existed during the accident about the pool water inventories. This was particularly true after hydrogen explosions (from hydrogen generated in the reactor cores and leaked into the secondary containments) caused extensive structural damage in three of the reactor buildings. Subsequently, pool water chemistry measurements confirmed that none of the fuel in the pools was damaged significantly and that none of the pools developed leaks even though the buildings were extensively damaged. But the location of these reactor pools remains problematic because the deposition of debris in the pools and the high elevation will complicate greatly the recovery of fuel from the pools. Thus, it may be preferable to move fuel from at least some types of spent fuel pools to on-site dry storage or possibly central interim storage on an expeditious basis.

The above discussion points to the need to re-examine the defense in depth of reactors to extreme external events and severe accidents. This will be most critical for currently operating plants since they have only had back-fit justified modifications as a result of the TMI2 accident, while the new ALWRs have largely already addressed these issues, with the passive plant designs having the greatest margin and resilience to such events.

Government Strategies to Support Innovation

In a well-known joke about innovation in nuclear energy technology, a utility executive explains that his goal in building new nuclear power plants is to be "number two." Nuclear power, in fact, is unique among energy technologies due to the substantial disincentives the technology provides for first-movers. These disincentives arise due to the long time periods and large capital investments that must be placed at risk to develop new reactor and fuel cycle technologies and the substantial difficulty in protecting the intellectual property that arises from taking risks. In particular, NRC regulatory decisions cannot be patented, so first-movers are faced with the fact that other vendors can "free-ride" on the knowledge gained if the first-mover is ultimately successful in obtaining NRC Design Certification, while the first-mover can lose its entire investment in development if the NRC decision is negative or the design is never deployed. Given the very large financial investment required to develop a new reactor design to the point where NRC licensing decisions are made, the fact that the first-mover places its entire investment at risk, while subsequent free-riders have greater confidence about likely regulatory decisions, creates a large barrier to first-mover investment.

It is a legitimate policy for the federal government to intervene in cases where market failures occur. Because free-rider problems are a well-known type of market failure, options exist for government policies to reduce these barriers sufficiently that socially optimal investment starts to occur in nuclear technology innovation. Worldwide, two major models exist by which national governments have supported the development of nuclear energy technology. They are discussed further below.

1 Innovation Emerging from Nationally Owned Reactor Vendors that Serve Captive Domestic Markets

Examples of countries where national reactor vendors have captive domestic markets include France, South Korea, and Russia.[2] China has imported reactors, but in the future it intends to rely fully on domestic reactor vendors as it masters the technology transferred from Western suppliers like Westinghouse. The vendors who have captive domestic markets are typically funded by their governments in the development phase of a new design and can recover first-of-a-kind demonstration costs from domestic consumers; in principle, they can compete in the export market based upon nth-of-a-kind (NOAK) costs. It appears difficult to achieve NOAK costs in exporting stick-build designs. For example, site labor has been one of the problems causing schedule delays and cost overruns for reactor designs like AREVA's EPR that use lower levels of modularity in construction. But NOAK costs may be achievable in exporting highly modular designs which have received additional engineering effort (and therefore cost) preparing the details ahead of an actual construction project.

However, it is not clear that the vendors that have captive domestic markets are capable of introducing substantive innovation into reactor design, since nationally owned reactor vendors tend to have more heavily bureaucratic management. For example, none have developed passive safety systems for domestic ALWR designs, although China would likely base future export reactors on the passive AP1000 technology it has acquired from Westinghouse. This type of environment is typically risk-averse and without precedent would be unwilling to tackle the challenges of significant innovation.

Another aspect that should not be ignored is the fact that such "nationally owned" reactor vendors get substantial commercial support in their bids into new markets. This support includes risk money for warranties, contingencies, and liquidated damages. In addition, political pressure and various side deals (e.g., for military aircraft) can also influence the outcome of reactor contracts in new markets as was the case with the recent successful bid by a South Korean consortium to build four reactors in the United Arab Emirates.[3]

2 Innovation Emerging from U.S. Federal Sharing of FOAK Development Costs

Several policy tools exist to address the free-rider market failure that discourages innovation in the development of nuclear energy technology. First, industry groups can form consortia to share FOAK risks. An excellent example is the DOE Nuclear Power 2010 program that, beginning in 2002, supported a consortium of eight U.S. utilities to develop Early Site Licenses (ESLs) for three reactor sites. In this program the focus was on ALWRs, which had strong customer support. NRC issued the first ESL developed under this program in 2007, and the utilities involved in this consortium have become the primary U.S. customers for new reactors.

More recently, the American Nuclear Society created a Special Committee on Small and Medium Sized Reactor Generic Licensing Issues, where representatives of several reactor vendors and developers, as well as independent experts, collaborated to develop a set of white papers addressing key licensing issues for such reactors.[4] The Reactor and Fuel

Cycle Technology Subcommittee of the Blue Ribbon Commission on America's Nuclear Future (BRC) has recommended that 5 to 10 percent of the federal nuclear energy R&D budget be directed to the NRC to accelerate the development of regulatory frameworks for advanced reactor and fuel cycle technologies and to perform associated anticipatory research.[5] The subcommittee noted that "[a]n increased degree of confidence that new systems can be successfully licensed is important for lowering barriers to commercial investment."

While creating industry consortia can help to reduce free-rider market failure, the fact that reactor vendors compete means that cooperative efforts can only work in the very early phases of the development of new reactor technologies, as with the ANS Special Committee. Eventually though, as in bicycle racing, individual reactor vendors must choose to break away from the *peloton* and compete individually to develop, license, and demonstrate a new reactor. Here the most effective policy tool is federal cost-sharing of FOAK engineering and NRC design certification costs to support detailed design and licensing efforts, plus federal loan guarantees, production tax credits, and purchase contracts for power and testing services to support demonstration efforts. In the case of small reactor development, the DOE has formulated a collaborative program with industry, similar to the Nuclear Power 2010 program, and has obtained congressional appropriations for fiscal year 2012.

The AP1000 provides an existence proof that these federal policies can be successful in overcoming first-mover barriers to the development and commercial deployment of innovative reactor technologies. The key element of these federal policy tools is that commercial entities have a strong self-interest to only pursue reactor development projects that have some plausible future commercial potential within a competitive market. Conversely, reactor vendors that have captive domestic markets are guaranteed a return on investment regardless of whether their technologies can compete internationally. For reactor vendors who face competitive markets, as U.S.-based vendors do, the decision to place some of their own capital at risk provides evidence that the vendors believe the technology has the potential to be commercially competitive.

The use of fifty-fifty cost sharing may not be appropriate in many cases. In the early phases of research and development, when risks are greater, a larger federal share is appropriate. At the stage of demonstration, however, production tax credits and purchase contracts for power and testing services can create incentives that encourage and reward effective construction project management. Also important for cost-share programs is recognition when there is "market pull" by potential customers. This can be gauged if these potential customers are willing to lend their reputations to the success of the program and financially support the program at a credible level. Federal efforts to encourage potential reactor customers to form consortia to develop ESLs proved to be very effective with the DOE's Nuclear Power 2010 program, leading to the first new nuclear construction in the United States in thirty years with two new plants each in Georgia and South Carolina. A new program could be very helpful in developing a new customer base for the use of small modular reactors to provide smaller increments of electricity as well as co-generation and process heat services to chemical and biofuels facilities.

Opportunities for Innovation

Three prime areas of opportunity for technical innovation are in fuel supply and waste generation, alternative nuclear energy products, and reduction of construction costs and schedules. The issues, and opportunities, for innovation in these areas are reviewed here.

1 Innovation to Address Nuclear Fuel Supply and/or Waste Disposal Costs

Experts generally expect that uranium prices will remain relatively low for many decades into the future, so that fuel supply costs will remain a second-order consideration in selecting reactor technologies for commercial deployment. Great uncertainty exists about what the ultimate costs of waste disposal will be, due to the current lack of established, large-scale geologic disposal capability for spent fuel and high-level waste and the great difficulty to date in developing these capabilities. But projected costs for geologic disposal of spent fuel remain sufficiently low (substantively lower than fuel supply costs) that waste generation also will remain a second-order consideration in selecting reactor technologies, unless governments mandate the use of different reactor technologies for waste reduction purposes (which thus far has not happened, although the idea that such mandates might occur provides the principal motivation for research on fast-burner reactors).

The fact that some type of geologic disposal will be required for all plausible future fuel cycles means that geologic disposal will ultimately need to be developed regardless of which future reactor technologies are deployed. A commercial risk for reactors designed to transmute, or convert, long-lived transuranic isotopes to short-lived isotopes (if such were deployed) is that prices for transmutation services could collapse rapidly due to expansion of geologic disposal capacity. The most extensively studied technologies for this transmutation mission are all based on fast neutron spectra. Past experience has shown that such reactors can have high cost and poor operating performance. Because of this and the fact that disposal of high-level wastes is by law the responsibility of the federal government, it is appropriate for the DOE to engage in "goal and schedule"-oriented R&D to determine the feasibility and cost of such reactors and their associated fuel cycles.

2 Innovation to Address Markets for Alternative Nuclear Energy Products Other than Large-Scale Commodity Production of Base-Load Electricity

Alternative markets for nuclear energy require smaller reactors, designs that can justify smaller plant site and emergency planning zone size, and typically higher operating temperatures. The potential markets include process heat for chemical production and industrial applications, desalination, space and naval power, and a variety of specialty applications for small-scale production of electricity. In general, large ALWRs cannot serve these alternative markets. While these alternative markets are unlikely to expand to the scale of commodity base-load electricity generation, they could emerge as important market niches where smaller reactors are required, providing opportunities where new innovative technologies can be deployed (e.g., LWR-based small modular reactors and Generation IV reactors).

While the commodity cost of base-load electricity will remain tied to the cost of coal, or natural gas for the cases where combined cycle gas turbine generation is the technology of choice regionally (at least as long as carbon emission costs remain low), natural gas and oil are currently the dominant fuels for these alternative applications. Because natural gas and oil prices are higher than coal prices, reactors designed for these niche markets can have somewhat higher production costs than would be acceptable for commodity base-load electricity generation. Currently, natural gas prices are low in the United States due to the expected expansion of shale-gas production, but chemical industry experts do not expect this to persist because sustained low natural gas prices will drive demand in existing natural gas markets and the deployment of gas-to-liquid conversion infrastructure that can arbitrage the price difference between natural gas and the significantly higher oil prices that are anticipated going forward.

A key activity the federal government could undertake would be to develop a "risk-informed" and "performance-based" regulatory framework for non-LWR technologies.[6] As discussed above, the BRC has recommended that 5 to 10 percent of the federal nuclear energy research, development, and demonstration (RD&D) budget be devoted to this purpose. The development of a risk-informed and performance-based regulatory framework is the only way that such technologies can be deployed on a reasonable schedule. This type of framework could give credit to the low risk of core damage/radioactivity release and allow co-location near process facilities. In addition, the federal government should cost-share the development of two to three NRC early site permit licenses for small reactors co-located to provide process heat to chemical facilities with the Next Generation Nuclear Plant (NGNP) consortium that was recently formed by a group of petrochemical companies and nuclear suppliers.

As already mentioned, a second key activity is to cost-share FOAK engineering and Design Certification costs for the development of LWR-based small modular reactors (SMRs) and to provide incentives to support the construction of demonstration plants, as these reactors could be attractive for co-generation of electricity and process steam. In the longer term, an alternative energy products market could provide an entry point for Generation IV high temperature reactor (HTR) technologies, particularly if many of the major regulatory issues for this market are addressed first for LWR-based SMRs. The availability of a market segment that would require smaller reactor size and pay some premium over commodity electricity generation prices could be helpful in reducing the future financing challenges for Generation IV reactor demonstrations, since all reactor technologies that use new coolants and fuels must be deployed initially with smaller reactor sizes.

3 Innovation to Reduce Reactor Construction Costs and Schedules While Maintaining High Reactor Safety and Operational Reliability

This last category is likely to be where most near and intermediate term innovation occurs. All major reactor vendors will be seeking approaches to reduce costs and construction schedules for large ALWRs. Further improvement can be expected for AP1000-derived reactors, but there will remain significant financing and logistical challenges for constructing large ALWRs that involve substantial design changes from previous reactors. For example, it is not yet clear whether the natural-circulation economic simplified boiling water reactor (ESBWR) will be demonstrated, even though it will receive NRC design certification.

There are also reasons to expect that some significant near-term commercial markets can emerge for LWR-based SMRs, due to their much easier project financing, the economies associated with repetitive factory manufacturing, and the potential for implementing design innovations that reduce, at least partially, the normal impact of economies of scale for reactors.

Finally, there is the entire topic of the potential for future Generation IV reactor technologies (that would use coolants other than water) to compete economically with LWR technology. Generation IV technologies face major development challenges, but also hold promise for improved performance compared to water-cooled reactors. This type of R&D, which can be characterized as very high risk but potentially high benefit, is a prime example of what the federal government should be funding in its entirety. The topic of Generation IV technologies and the possible pathways by which commercially successful Generation IV reactors might emerge (none have to date) is important, and thus is discussed separately in the next section of this paper.

Issues for Generation IV Technologies

The Generation IV International Forum (GIF) was founded in 2001 as a cooperative international endeavor to carry out the R&D needed to establish the feasibility and performance capabilities of the next generation of nuclear systems. The attributes that these next-generation nuclear systems seek to achieve include improved economics, enhanced safety and reliability, sustainability (from fuel utilization and waste minimization standpoints), and greater proliferation resistance and physical protection. A roadmap completed by the GIF in 2002 identified six categories of Generation IV designs that differ significantly from LWRs operating today or being developed:

- very high temperature reactor (VHTR, graphite moderated and gas cooled),
- sodium fast reactor (SFR),
- lead fast reactor (LFR),
- gas-cooled fast reactor (GFR),
- molten salt reactor (MSR), and
- supercritical water reactor (SCWR)

All of these designs, and various potential derivatives of these designs, have significant technical, regulatory, and economic challenges to overcome before they can compete commercially with the well-understood and developed ALWR technology available today.

Challenges in Developing Generation IV Technology

Because Generation IV reactors use coolants and fuels that differ greatly from ALWRs, materials issues create some of the fundamental challenges to be overcome due to the high operating temperatures and potentially more aggressive coolants that are part and parcel to these designs. These designs present new and different safety issues (e.g., sodium fires, air ingress oxidation, etc.) that are not associated with LWRs. While these designs may eliminate many of the specific safety issues posed by LWRs, the new safety issues they create are not well addressed in the current NRC regulatory framework. No systematic approach exists to credit these systems for the safety issues they eliminate and to assess their overall safety compared to ALWRs.

Probably the most important problem from a deployment standpoint is the very uncertain economics of these concepts once designed and licensed. Given the very long time periods typically needed to develop and qualify new materials and fuels, pragmatic decisions often must be made in the design of commercial-scale demonstration reactors to select more conventional, lower-performance materials, to reduce operating temperatures, and to de-rate fuels and power levels. As a result, first-generation designs can have significantly higher construction and operating costs than might be achieved once the technology has fully evolved. Past experience has been poor in the Western world from capital cost and operational reliability standpoints. The capacity factors of commercial-scale demonstration high temperature reactors (e.g., Fort St. Vrain in Colorado and a thorium high temperature reactor in Germany) as well as the sodium fast reactors (e.g., Superphénix in France and Monju in Japan) have been very poor. These reactors have been plagued by maintenance and operation problems that resulted in very low capacity factors, which in many cases have resulted in permanent plant shutdowns.

Licensing Basis Events: Predicting Frequency and Consequences

The challenges for developing, licensing, and commercializing Generation IV reactor technologies can be understood by reviewing the "Farmer's chart" shown in Figure 7.5. The Farmer's chart is a convenient tool to display the licensing basis events (LBEs) that have been considered in the design of a Generation IV reactor like the pebble bed modular reactor (PBMR). LBEs are identified from the development of event trees using a systematic process that considers "internally" initiated events, caused by combinations of equipment failures and human errors, and "externally" initiated events arising from sources like earthquakes and floods.

LBEs are categorized by their frequency, with anticipated operational occurrences (AOOs) being events with frequencies greater than once

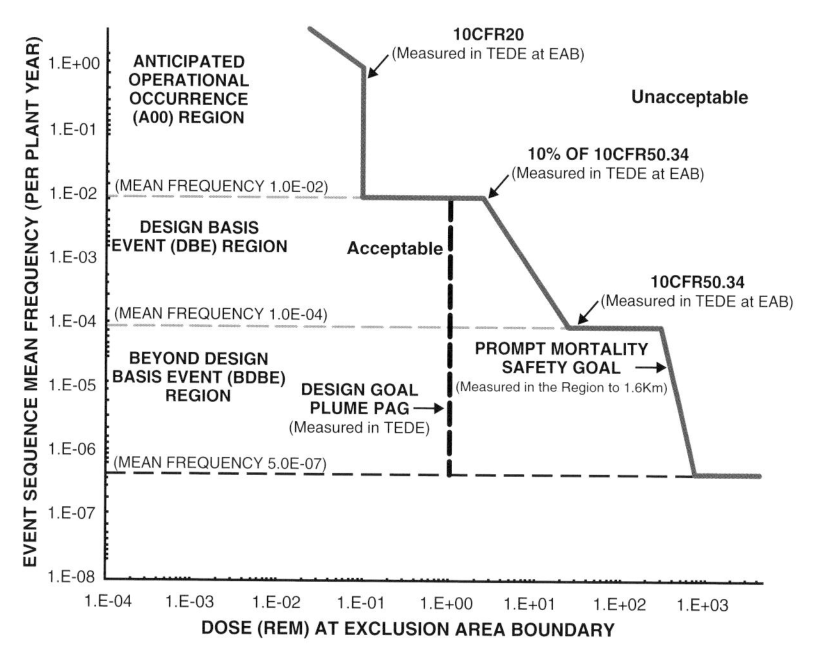

Figure 7.5 A Farmer's chart showing licensing basis event frequencies and consequences. Figure courtesy of the Department of Energy's Idaho National Laboratory, run by Battelle Energy Alliance.[7]

every 100 years, that can reasonably be expected to occur during the life of a plant; design basis events (DBEs) having frequencies of every 100 to 10,000 years, that could reasonably be expected to occur at least during the life of a fleet of plants; and beyond-design basis events (BDBEs) that are unlikely to occur even in a fleet of plants but are designed to protect public health and safety.

The normal metric used to assess the consequence of LBEs is the maximum radiation exposure to a member of the public living at the plant site boundary, expressed as a total effective dose equivalent (TEDE) that combines both internal and external radiation exposures. In the Farmer's chart, this dose appears on the horizontal axis.

Following an LBE, the transient phenomena that can result in damage to the reactor, release of radioactive material, transport to the site boundary, and exposure to the public can occur over a relatively short time period of minutes to days. To license a reactor design, a vendor must develop and experimentally validate transient modeling tools to predict damage to reactor systems and fuel, release of radioactive material, and transport and exposure off site. For the licensing of earlier reactor designs, relatively simple, deterministic modeling methods were used that were aimed at demonstrating that compliance with regulatory release limits could be achieved, but that did not attempt to predict a best estimate of the actual damage or release, or the uncertainty in these predictions.

The best-estimate modeling methods that have been developed and applied to the licensing of ALWRs with passive safety systems, such as the AP1000, can also be applied to Generation IV systems. Adapting these best-estimate modeling methods to Generation IV technologies will require effort and competence. The effort is merited because the modern best-estimate modeling methods provide a systematic and rigorous approach to demonstrate that models have been appropriately validated with experimental data and to quantify uncertainty in the model predictions so that safety margins can be demonstrated. The structured approach of best-estimate modeling for reactor transients can make it possible for the NRC to assess the modeling methods and supporting experimental data at an early phase of the development effort for a new reactor design (during pre-application licensing review), which can then increase confidence for the reactor vendor that results generated by these models will provide an adequate basis for licensing the unique safety features used in their design, before a large investment must be made in detailed FOAK design engineering.

A more fundamental problem for Generation IV systems is the need to predict the frequency of initiating events, the vertical axis on the Farmer's chart. ALWR designers can rely upon the massive database of LWR operating experience and component reliability data that have accumulated during many thousands of reactor-years of LWR operation. ALWR vendors can therefore be strategic in taking on limited and well-defined risks from using new materials and component designs to maximize the incremental benefits from these changes. For example, the AP1000 is the first large pressurized reactor design to develop, qualify, and use "canned" primary pumps. Canned pumps place the motor rotors inside the primary pressure boundary and thus eliminate the need for high-pressure shaft seals and associated seal leakage. However, there is a large body of operating experience of such pumps in naval reactor programs (albeit at much smaller sizes) that gives confidence to their reliable operation.

Because Generation IV reactors, of necessity, use a large number of systems, structures, and components that are novel and for which relatively little operational experience and reliability data exist, they can require large development efforts to test and qualify materials, fuels, and components. Besides being expensive, these efforts can require many years and even decades of investment and work because typical fuels must survive from four to as long as thirty years of irradiation to reach full burn-up; reactor components are often required to have service lives ranging from ten to sixty years. Because unanticipated component failures have been a principal cause of low plant availability in previous Generation IV reactor demonstrations, successful future demonstration programs will require a well formulated and executed program of materials, fuels, and component testing, combined with rigorous application of reliability engineering principles in system design (to identify materials degradation mechanisms, to identify resulting component failure modes and effects, to design instrumentation for detecting and monitoring degradation progression and effects, and to design components to facilitate maintenance and replacement).

What Path Forward for Generation IV?

The challenges to developing commercially competitive Generation IV reactors are daunting. While some Generation IV reactors can use uranium more efficiently and reduce the amount of waste requiring geologic disposal, the commercial benefits of reduced uranium and waste disposal costs are likely to remain small for the foreseeable future because uranium and waste disposal currently contribute less than 10 percent to the total cost of power from new nuclear plants. This said, it is difficult to imagine that LWRs will remain the economically optimal technology for producing nuclear power in perpetuity. The interesting question is by what path the first Generation IV reactor technology might emerge with the ability to compete with, and ultimately displace, LWR technology in competitive energy markets.

While all of the six Generation IV technologies identified by the GIF in 2002 have been studied using test reactors (except the supercritical water reactor, a high-temperature, high pressure variant of the LWR), to date none have been extrapolated to commercially successful prototype designs. This raises the interesting question of whether the earliest commercially attractive Generation IV technologies might emerge from a hybridization of different Generation IV technologies. A hybridized design with the capability to compete commercially with LWRs could provide a base of operational experience, component reliability data, and design and licensing tools needed to evolve toward subsequent advanced Generation IV designs that would enhance the specific strengths of the six original concepts—the higher temperatures of the VHTR, the capacity of the fast-neutron spectrum designs to transmute wastes, and/or the simplicity of spent fuel recycle and fuel fabrication of the MSR.

With this premise in mind, the University of California-Berkeley has studied a variety of potential hybrid high temperature reactor designs since 2002 in collaboration with Oak Ridge National Laboratory, the University of Wisconsin, and the Massachusetts Institute of Technology.

The option for a hybridized Generation IV reactor that has received the most study would use the high temperature fuel developed for the helium-cooled with the high temperature liquid fluoride salts studied earlier for the MSR, resulting in an advanced high temperature reactor (AHTR).[8]

The high heat transport capacity of fluoride salts, which exceeds that of water, allows AHTR reactor cores to operate at power densities four to six times greater than helium-cooled VHTRs with the same fuel. Besides resulting in a much more compact reactor design than the VHTR, the AHTR fuel also reached full burn-up much more rapidly than VHTR fuel, typically in less than one year of residence in an AHTR core. This means that the time required to develop, test, and qualify new fuels from AHTRs is far shorter than for other Generation IV reactor designs.

While the specific path that will result in commercially competitive Generation IV reactor technology remains uncertain, key elements of this path can be hypothesized. First, investment risk can be reduced if best-estimate transient analysis codes are developed early on, validated with well-designed experiments, and assessed by the NRC during the pre-application review phase. Second, the selection of the combination of fuel, coolant, structural materials, and power conversion technology must be strategic and must be supported by a well-formulated development plan for materials and component testing. Finally, federal policies must create an environment that enables the development, detailed design, licensing, and demonstration of commercially competitive Generation IV technology. Key ingredients include:

1. Continuing a strong federal and industry (EPRI) base program of R&D in materials, fuels, safety, separations, and waste forms, supported by experimental capabilities including test reactors for irradiation and facilities for post-irradiation examination and chemical separations.
2. Increasing the investment in NRC anticipatory research and regulatory framework development to be 5 to 10 percent of the total federal nuclear energy R&D budget.

3. Supporting a consortium of utilities and chemical companies to develop two to three early site permits for SMRs co-located with chemical and low-carbon transportation fuel facilities to provide process heat and supporting two reactor vendors to perform FOAK engineering and design certification of LWR-based SMRs that could be used for early process heat and co-generation at these sites. This would create a new market that would require smaller reactors. Reactor vendors who are interested in demonstrating smaller Generation IV reactors (and thus place less capital at risk in the demonstration project) could then anticipate selling additional small reactor units for these new process heat and co-generation applications that require higher operating temperatures.
4. Developing a graded structure for sharing FOAK risks that increases the federal share in the early phases of development, design, and licensing, and reduces the federal share at the time of demonstration, for Generation IV reactor development, instead of using a simple fifty-fifty cost sharing rule. Federal incentives for the construction of FOAK demonstration reactors should be weighted to encourage competent project management and schedule adherence. One way would be by emphasizing the use of incentives that reward successful completion and operation of reactors, such as power and test-service purchase contracts and production tax credits.

Commercial interest in developing and demonstrating Generation IV reactor technologies remains stalled, while commercial deployment of conventional large ALWRs is moving forward and substantial commercial interest exists to develop LWR-based SMRs. Innovation is a key ingredient that can change this balance in the future and lead toward a long-term transition away from water as the dominant reactor coolant.

Conclusions

Nuclear energy is a challenging and important technology. While innovation in nuclear technology involves substantial risks, innovation is also a vital element for commercial competitiveness. The DOE ALWR and Nuclear Power 2010 programs—which cost-shared the development of early site licenses with the NuStart consortium of utilities and the design certification of two ALWR designs—were essential contributors to the resumption of construction of new reactors in the United States. Further improvements in nuclear energy technology—to further increase safety, reduce construction costs, increase efficiency, expand products to include process heat and desalination, and reduce waste generation—are possible, but have very significant deployment hurdles. Federal policies to support innovation, as described in this paper, can accelerate this progress.

Notes

1. The distinction between active and passive safety is an important one. The concepts of active and passive safety describe the manner in which engineered safety systems, structures, or components function and are distinguished from each other by whether there is any reliance on external mechanical and/or electrical power, signals, or forces. See *Safety related terms for advanced nuclear plants*, IAEA-TECDOC-626 (Vienna: International Atomic Energy Agency, September 1991), pp. 1–20.

2. Both France and South Korea initially relied upon reactor technology developed in the United States before they developed indigenous capabilities.

3. *The Economist*, "Unexpected reaction: The handful of firms that build nuclear reactors face new competition," February 4, 2010, http://www.economist.com/node/15457220.

4. The ANS Special Committee's July 2010 interim report can be found at www.ans.org/pi/smr/ans-smr-report.pdf.

5. The commission was established by Secretary of Energy Steven Chu, acting at the direction of President Obama in January 2010, to conduct a com-

prehensive review of policies for managing the back end of the fuel cycle (more can be found at http://www.brc.gov).

6. The term "risk-informed regulation" refers to examining the likelihood of an event occurring in addition to its possible consequences when assessing risk. A risk-informed approach contrasts with a solely "deterministic" approach to regulation which can lead to unnecessary requirements. The term "performance-based" refers to regulations that focus on desired, measurable outcomes and contrasts with "prescriptive" regulation where actions are prescribed by the regulator.

7. "Next Generation Nuclear Plant Licensing Basis Event White Paper," Idaho National Laboratory, INL/EXT-10-19521, September 2010, page 20.

8. Charles W. Forsberg, Paul S. Pickard, and Per F. Peterson, "Molten-Salt-Cooled Advanced High-Temperature Reactor for Production of Hydrogen and Electricity," *Nuclear Technology,* vol. 144, no. 3 (December 2003), pp. 289–302.

The Spent Fuel Problem

8

ROBERT J. BUDNITZ

Introduction and Background

The issue of how ultimately to dispose permanently of the high-level radioactive wastes generated by the use of nuclear power, and also generated during the manufacture of nuclear weapons, has been a contentious one for decades. The related issue of how to store this dangerous material in the "interim" before its ultimate disposal has also been contentious, especially because with no ultimate disposition path on the immediate horizon, the term "interim" means at least a couple of decades from now, and perhaps longer. In both cases, disposal and interim storage, the issues are political, not technical. It seems not to be widely known that we do have an operating repository in New Mexico for one category of radioactive waste, the defense program's transuranic wastes, called the Waste Isolation Pilot Project. WIPP was developed through a process involving the state and the locality. The Obama administration's Blue

Ribbon Commission (BRC), set up to recommend how to proceed with the much larger amount of power plant waste, lays out the problem and proposes a solution to the site issue and organizational issues.

Why Deep Geological Disposal?

There is very broad agreement that the material at issue should not be disposed of permanently near the surface, such as in shallow land burial, or even in engineered facilities at or near the surface. A large number of careful studies and reviews, both domestically and internationally, and going back a half century or more, have all concluded that for the very long term (meaning for millennia or even millions of years) the only management approach that can provide adequate safety is deep underground disposal. The material simply poses too great a hazard, not only to human health but to the broader environment, to be disposed of on or near the surface, even in the best engineered facility that anyone can imagine deploying today or anytime soon.

What Is the Material?

The material at issue is mostly used nuclear reactor fuel, containing radioactive fission products and actinides. About 60,000 tons of this used fuel is already stored in the United States, and about 2,000 tons arise annually from the 104 operating U.S. power reactors. Abroad, the total is about four times larger. In the United States, it is essentially all still located at the sites where it was generated by the reactors. Its composition varies from reactor to reactor, although these compositions are broadly similar. Most of the used reactor fuel comes from commercial power plants (all of which in the United States today are LWRs, light-water reactors). But there is also used fuel from naval-propulsion reactors and various research, test, and isotope-production reactors. There is also the waste from the U.S. nuclear-weapons program, a lot of which is still in waste tanks (although most of the tank wastes have now been dried out). Other defense wastes are in various other forms (such as glass) that have been prepared for ultimate disposal.

Safety and Security during Interim Storage

The wastes under discussion here are essentially all being stored at the sites where they were generated. All storage of used commercial reactor fuel must be licensed by the U.S. Nuclear Regulatory Commission (NRC) and all storage by the government (principally by the Department of Energy and the Navy) must follow comparable requirements. When used fuel is initially discharged from a power reactor, it generates so much (thermal) heat that, if not cooled well and continually, the heat would soon melt the fuel rod cladding, thereby releasing the enclosed radioactivity. Therefore, this fuel must initially be kept in a *fuel pool* (under water) with active systems to remove the heat from the water and keep the water level up. This arrangement is necessary for three to four years, after which the heat generation has decreased enough to allow the used fuel to be removed from pool storage and placed in so-called *dry-cask storage*. However, in the United States today, this transfer to dry-cask storage typically does not take place until at least ten years later, and some used fuel has remained in pool storage for up to three decades.

Technical Consensus about the Efficacy of Deep (Repository) Disposal

There is a broad scientific consensus today, shared almost universally, that it is technically feasible to dispose safely of this highly radioactive material deep underground in a mined repository, if suitably designed. In fact, the design of the Yucca Mountain repository, now in its death throes because of political opposition, is one such design. The NRC, charged with reviewing the design and ultimately granting a license for it, completed its design review in 2010. Although politics have intervened, the NRC staff finding, recently released in interim form by a congressional committee, is that no technical issues stand in the way of the Yucca Mountain repository design ultimately receiving a license to proceed.

Technical Consensus about the Safety and Security of Spent Fuel Storage

There is also a broad consensus, shared almost universally, that when done correctly the storage of spent fuel in pools is both safe and secure, and that the storage in dry casks is even more so. However, nobody thinks that dry casks represent more than an interim solution—they simply should not be relied on for millennia, because the casks must be rebuilt or the waste moved to new ones every century or so. That would mean relying on active institutions in perpetuity, which nobody thinks is a wise policy.

So What to Do?

Again, the broad consensus is that the failure to carry through with a program for deep geological (mined) disposal is *a failure of policy, not a failure of technology.*

To address this failure, the Obama administration appointed the BRC in early 2010 to examine this issue. The BRC's report, issued in February 2012, has identified a small set of important policy changes and approaches which the members believe can allow a sensible national waste-disposal program to proceed to a technically and politically acceptable endpoint.

A paraphrase of the most crucial of these BRC findings and recommendations is as follows.

1. Selecting a site for a radioactive waste repository must be a bottoms-up process in which the host community is involved from the start, concurs, and supports the repository. The Swedes and Finns have

used this approach successfully. The only operating repository in the United States, the WIPP site in southern New Mexico for disposal of certain defense radioactive wastes, is enjoying strong local support. Imposing a disposal site on a hostile host area, as Congress did with Yucca Mountain in 1987, is doomed ultimately to fail.

2. The Department of Energy experience with Yucca Mountain has shown that asking a politically engaged and controversial agency to carry out a repository development program, which must survive for decades over many different administration changes, is also unlikely to succeed. The Blue Ribbon Commission recommends founding a quasi-governmental entity modeled on the Tennessee Valley Authority, which it claims has the promise of being able to carry out what a cabinet department has shown it cannot. This approach would significantly de-politicize the issue.

3. The costs of the repository are, by current statute, borne by the nuclear plants and their ratepayers through a small tax or fee on the electricity they generate. This goes into a federal waste fund. The logic of this is impeccable. But many years ago, Congress took this apolitical fund and politicized it by making the process of appropriating expenditures from this fund a political part of the overall federal budget process. This gave Congress an incentive to seize parts of the fund to "reduce the deficit"—which it has done annually ever since, never even once appropriating what the Yucca Mountain repository program said it needed. As the Blue Ribbon Commission has recommended, the fund must be dedicated by statute to the repository and to nothing else, and must be available as needed to the TVA-like government entity charged with developing the repository.

4. Retrievability of the radioactive material disposed deep underground, at least for a century or so, is important. This is because there will inevitably be an evolution in technology in many areas related to radioactive waste and its repositories. To account for this, the current regulations governing Yucca Mountain require that the waste, once emplaced deep underground, must be retrievable for a

century, so that our grandchildren or their grandchildren can then decide whether to seal up the repository as a permanent disposal site or to retrieve the waste if, by then, it is found to have some beneficial use (or if the repository design is somehow judged by then to be inadequate). This is a sound policy.

5. In particular, with some foreseeable technological advances, it may turn out to be economically feasible and technically desirable to reprocess the waste (or some of it) chemically, so as to extract any useful constituents for beneficial use. This may include the plutonium in it that would otherwise be disposed of, but which might later be re-used as a fuel for advanced reactors—either for reactors to burn as fuel or for reactors to use to breed even more reactor fuel. This would extend our uranium fuel supply by a large factor—say, fifty or a hundred—to provide millennia of nuclear fission fuel.
6. Such reprocessing would also remove most of the very long-lived radioactive material from the waste stream, meaning that the disposal task for the rest would be much less challenging. Most of the remaining waste would require sequestering only for a few centuries rather than for tens of millennia or even longer, a far easier and less expensive engineering task and one offering an even higher assurance of success.
7. Security concerns with the waste must be considered. The broad consensus today is that the waste being stored in either fuel pools or dry casks is adequately secured against security threats, and that deep geological disposal would be even more secure. This is again not a difficult technical issue with appropriate attention, although as always it is a political or institutional task to assure that surface management of these wastes is done everywhere to the high standards it requires.

International Issues Related to Nuclear Energy

9

WILLIAM F. MARTIN AND BURTON RICHTER

Introduction

We are entering a new era of increasing demand for electricity, essential for economic growth in the developing world. Nuclear power is on the agenda of many of these nations, but there is a concern that they do not have the necessary legal and regulatory infrastructure to ensure safety and security. In some cases, they may also lack the will to dedicate enough financial and manpower resources to the endeavor. As we have learned once again from the Fukushima incident in Japan, a nuclear accident anywhere is an accident that will have implications globally.

In the wake of Fukushima, serious questions have arisen about the safety of reactor designs, emergency back-up systems, on-site spent fuel storage, and the regulatory systems governing nuclear power. These debates are occurring actively across the world. Ironically, the greatest impact is likely to be in nations which are members of the Organisation for Economic Co-operation and Development (OECD) as their publics raise legitimate concerns about nuclear safety and security, while the greatest danger is likely to be in countries that lack a tradition of open discussion.

Worldwide Growth of Energy Demand

Today there are over 440 nuclear reactors operating in thirty nations.[1] As noted in the chart below, most operating reactors are in technologically advanced nations with nuclear experience.[2]

But experience does not necessarily assure safety: we have witnessed the most severe nuclear accidents in the United States, what is now Ukraine, and Japan. These experiences underscore the importance of developing nuclear energy with a "safety first" attitude, especially in nations that are new to nuclear power. If one compares Figures 9.1 and 9.2, it is clear that the growth of nuclear power is likely to be in nations experiencing rapid growth and need for electricity, many of which are not yet advanced in terms of their governmental institutions.

Nations in Europe with a large and politically strong green agenda have been the ones most affected by the accident at Fukushima. The close proximity of nations that have turned against nuclear power (e.g., Germany)[3] and those that continue to support it (e.g., France) necessitates continued involvement of all European nations, regardless of their stance on nuclear power. In the United States, nuclear construction will slow due to lower electricity demand and gas prices. The U.S. electricity sector is likely to remain primarily fueled by coal, natural gas, and shale gas for some time.

While many industrialized countries pause for reflection, nuclear energy growth in industrializing nations will be rapid. The motivation for nuclear power in these countries arises from economic and political

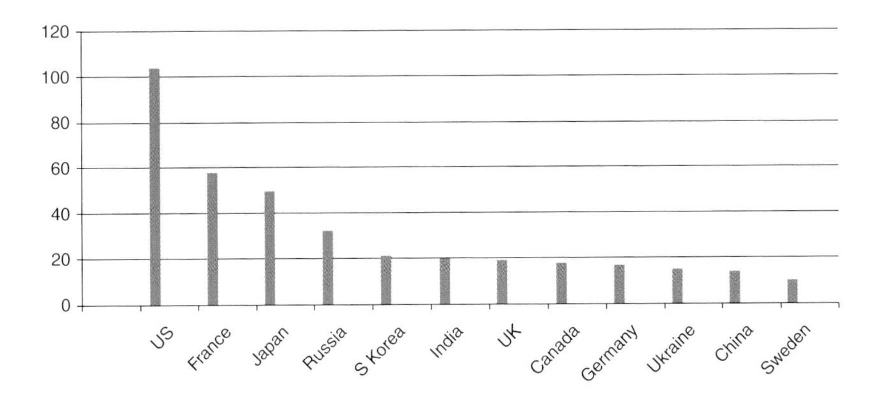

Figure 9.1 Existing Reactors Worldwide.[4]

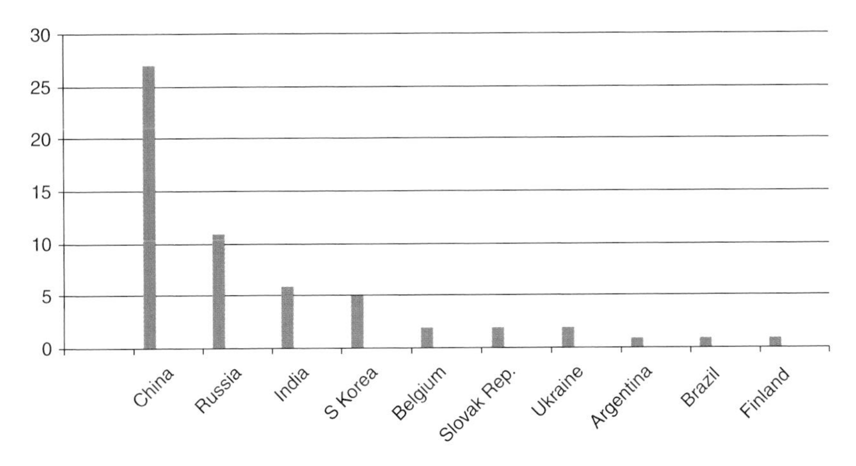

Figure 9.2 Nuclear Reactors Under Construction.[5]

factors, as well as the prestige of having nuclear technology. A major reason for the growth of energy demand in developing countries has been the inspiration of world leaders in the early 1980s to commit to a regime of enhanced international free trade, allowing manufacturing to move to the most efficient and cost-effective regions and countries. Thus, the steel industry moved from the United States to Japan to South Korea to China and so forth.

This shift in manufacturing resulted in higher energy/gross national product (GNP) ratios in industrializing countries, while OECD countries moved toward less energy-intensive service industries. Thus, it is no surprise that the areas of major energy growth of the last twenty years have been in China, India, Southeast Asia, and Latin America. Likewise, the world's growth of energy will be in population-intensive, low-cost manufacturing nations.

The end of the Cold War and this economic expansion together lifted much of the world out of poverty and totalitarianism. They also opened up the wish of economically developing nations to secure the nuclear energy option. Nuclear energy is viewed as advanced technology and, as such, carries a certain prestige as well as economic benefit in a world of fluctuating oil and gas prices. And, undeniably, it is a cleaner fuel than coal. The projections of the International Energy Agency, however, do show considerable coal growth.[6] In fact, coal may grow twice as fast as nuclear. Nevertheless, the reality is that we are likely to have at least 150 new nuclear plants in developing countries within the next twenty-five years. Even at that pace, the share of nuclear power is going to fall as a percentage of total electricity.

The issue is not the overall number of reactors globally, but the location of "new builds" of reactors. The 150 new reactors expected by 2035 will primarily be in nations beginning their nuclear energy programs; many of them are under construction now.

Much of the already existing growth (reactors either planned or currently being built) is in China, Russia, and India, which have huge projected future domestic power demand increases.[7] There are other nations that are also planning to expand their limited existing programs, such as Poland, Lithuania, Argentina, Brazil, the Czech Republic, and South Africa. There are also nations with no nuclear energy programs today that have indicated they wish to enter the field, such as Egypt, Ghana, Jordan, Indonesia, Malaysia, Nigeria, Saudi Arabia, the United Arab Emirates (UAE), Vietnam, and possibly others.

The projections we show, therefore, are not wildly optimistic; they are, in fact, likely levels of growth of nuclear power even in the post-Fukushima world. It is true that public reaction in OECD countries has dampened the enthusiasm for nuclear energy expansion, but this is not the case in developing nations, especially those nations in the Middle East and Asia that are autocratically and/or centrally ruled. In these nations, it is the will of the government, not necessarily the will of the people, which will determine nuclear energy deployment.

During the 1970s, Professor Alan Strout, an energy economist from the Energy Laboratory at the Massachusetts Institute of Technology, did some innovative work on consumer purchasing patterns at different levels of per-capita income. For example (in 1970 dollars), a consumer would be motivated to buy a small cooker at $100 per capita, a heater at $250 per capita, a refrigerator at $500 per capita, a car at $1,000 per capita, etc.[8] Imagine the purchasing power of a relatively youthful global population as it enters the workforce and begins families. Of course, the key to this is the jobless rate in developing countries and jobs are dependent on economic growth, which is dependent on consumer purchases. At the moment, the system is stalled somewhat. But once the cycle of economic recovery begins, energy growth is expected to expand rapidly. Again, this will not be in the "mature" economies of the OECD countries, but rather the energy-hungry world of developing nations as the world grows from 7 billion people to over 10 billion by 2100.

The reason this fundamental economic analysis is necessary is to deflate the opinion that industrializing countries are purchasing nuclear power plants solely to gain "nuclear technology." Certainly this is one aspect of the wish to have nuclear reactors, but it is also in many cases wise economics as fossil resources might rise in price.[9] In any event, it is clear that nuclear power will grow, perhaps rapidly, due to energy security and economic reasons and less so because of the concern over climate change.

Let's take China as an example. From a relatively small base, China plans to build up to a capacity of eighty gigawatts of nuclear electricity (GWe) in the next twenty years. Even so, nuclear power will meet only 5 percent of Chinese electricity needs by 2030.[10] So it is not an issue of *whether* nuclear reactors will be built in industrializing countries, it is a matter of *how many*. And as Figure 9.2 demonstrates, the building is already under way and the race is on by commercial vendors to supply the nuclear energy option in the Middle East, Southeast Asia, Latin America, and Eastern Europe. While the current global downtrend will slow this development somewhat, the growth potential is evident.

Criteria for Safe Nuclear Development

Measures to ensure safety, security, and nonproliferation are weak, in part due to the fractured nature of the nuclear industry and the patchwork of national and international standards aimed at ensuring safe deployment and adequate regulatory oversight. Figure 9.3 demonstrates the adequacy of the "nuclear infrastructure" in selected regions of the world. This is simplistic, but it shows that there is little consistency in approaches to safety and security. It was surprising that an accident the size of Fukushima could happen in Japan, a nation widely regarded as one of the most technically competent in the nuclear world. While the analysis of Fukushima is not yet complete, it is already clear that a contributing factor was the weakness and lack of independence of the regulators, just as that was a major contributing factor to the 2010 Deepwater Horizon oil blowout in the Gulf of Mexico. One can only imagine what

	Infrastructure	Regulation	Knowledge Base	Safety/Security	Nonproliferation
OECD					
E Europe					
China					
Russia					
India					
SE Asia					
L. America					
Middle East					

The above table uses the color coding from best (dark blue), to OK (pale blue), worrisome (yellow), and lacking (red).

Figure 9.3 Where Nations Stand.

would have happened in such a situation in a less technically competent country.

Learning the lessons of Fukushima will be important. But applying them to existing nuclear energy nations—as well as to those that are seeking to develop nuclear programs—will not be simple. The reality is that the growth of nuclear power will have to be accompanied by an unprecedented effort to ensure high safety standards and adequate independent regulatory authorities. A talent pool to operate nuclear power safely will also be needed. Some nations, such as those in Eastern Europe, have the operating talent but may lack regulatory talent. Russia, Japan, and France may have the needed talent pool, but their regulators are not totally independent. We are fortunate in the United States that we have a rigorous Nuclear Regulatory Commission (NRC). But that only came after learning the lessons of Three Mile Island (TMI), lessons that took decades to learn and apply in a modern society with a huge talent pool to draw upon from leading universities and national laboratories.

Another challenge of the nuclear world is the ferocious competition among vendors to sell nuclear reactors and services, often led by governments that are seeking international markets to deploy their nuclear energy reactors and systems. As was the case with the recent reactor sale to the UAE by South Korea, the most economically competitive price is going to be the strongest inducement to make the purchase.[11] Burdening competitive bids with safety options that add to cost is not likely to be successful. The world is seeking the Volkswagen of reactors, not the Mercedes or Lexus. One possible route for worldwide nuclear power growth is to encourage the vendor to put together an entire package of measures to build, implement, train, and operate nuclear plants, including take-back of fuel—basically, a turnkey operation. This would ensure a safety-first approach if coupled with international standards that would be published by the International Atomic Energy Agency (IAEA) and adopted universally by nations new to nuclear power as well as those that already have reactors.

Another issue is that the nations that have the highest safety and security standards (France, Japan, and the United States) also have complicated export control regimes. The purpose, of course, is to ensure adequate safeguards. But the result is that potential buyers may seek the less costly reactors that are built to lower standards. From the point of view of the United States, this has the practical implication of driving business to competitors who have less rigorous standards. A key element of international cooperation will be the need to improve the regime of technology transfer to a level comparable to the "CoCom" regime during the Cold War.[12] In a bilateral relationship, the United States is wise to insist that a nation commit itself to higher standards. However, as the United States loses the commercial edge in international deals, its political power will be reduced.

An interesting recent debate in the Nuclear Suppliers Group (NSG) resulted in a decision to refuse to ship sensitive technology to India despite assurances in the U.S.-India nuclear deal. New NSG guidelines would make transferring such technologies more difficult. But it is interesting to note that France is aggressively pushing ahead regardless, as it sees the Indian commercial nuclear market as a huge opportunity.[13]

Complicating the situation is the debate that is emerging in forums like the IAEA. There is a well-known balance that IAEA needs to achieve given its diverse membership. In essence, the grand bargain is that the Group of 77[14] seeks nuclear technology (not only in reactors/fuel cycle technology, but nuclear medicine and other non-electricity nuclear technologies) as a payout for support for nonproliferation, safety, and security policies sought by the United States and other OECD/Nuclear Energy Agency (NEA) nations. A recent review of the IAEA's 2020 Report[15] revealed that this grand bargain is necessary to achieve a safe and secure nuclear future.

The legal reach of the IAEA, however, only extends to safeguards; it does not include safety and security. The IAEA can only advise on these functions. During the recent Fukushima accident, the IAEA had to have its Fukushima reporting approved by the Japanese government prior to release, which led some to criticize the IAEA for its slow response. Today, the IAEA is a watchdog for nonproliferation, but not for safety. However, there is hope that new mechanisms will be forthcoming. In its July 2011 letter to IAEA Director General Yukiya Amano, the International Nuclear Safety Group outlined some of the steps necessary to strengthen the IAEA in this regard, including: increasing world safety standards; expanding peer review services; updating current conventions; and expanding international emergency response and preparedness.[16]

Key issues:

- Infrastructure and regulation are the keys to a successful and safe nuclear program.[17] Of special importance is the independence of the regulatory body.[18] The events at Fukushima demonstrated the need for a strong independent regulatory system, even in technologically advanced nations. It must be remembered that the NRC, today considered the gold standard, was strengthened because of the lessons learned from TMI.
- A strong knowledge base and human capital infrastructure are also necessary for the long-term sustainability of a nuclear program. This includes university programs, research programs, and a large pool of workers.[19] In many developing and newcomer nations, these do not exist. Even in developed nuclear nations, years of neglect and underfunding, coupled with an upcoming wave of retirements across all nuclear fields, will exacerbate the situation.
- Safety and security are essential parts of any nuclear program because, as we have learned from Fukushima, an accident anywhere is an accident everywhere for nuclear. Without a strong commitment to safety among all nations, worldwide nuclear power growth will be in jeopardy. Without public faith and confidence, especially in developed nuclear nations, nuclear power cannot advance.
- Nonproliferation concerns must also be considered in future technology-sharing agreements and commercial cooperation efforts. This will factor heavily in any government-to-government deal; in fact, it already does, as exporting nations (primarily the United States) are very concerned about the nonproliferation risks in newcomer nations. There is also a danger of increased proliferation risk because nonproliferation concerns may be trumped by commercial competition among vendors, especially as many vendors are either state-backed or state-owned. The nuclear supply

chain is global and the United States is only one of many supplier nations.

There are many areas of concern when considering nuclear power. Figure 9.3 on page 214 displays which nations and regions are best positioned on each issue.

Priorities for Safety

Safety is the most important aspect for nuclear power in all nations with nuclear reactors. All nations should be evaluating current practices in light of the events at Fukushima. Not only developing nations and newcomers should be increasing safety protocols, as the events of Fukushima have demonstrated. In order to ensure safety in global nuclear operations, certain priorities must be set, including the following.

- Energy policy planning and nuclear energy: In making future energy policy decisions, nations (especially newcomer nations) must factor in all costs associated with nuclear energy, not just the "cash price" for the actual reactor system from the foreign vendor.[20]
- Independent regulatory authority: As we have learned from Fukushima, an independent regulatory authority is a must, not only for newcomer or developing nations but for all nations that utilize nuclear power.[21]
- Location, construction, and grid connection: When determining locations of new reactors, potential natural disasters (e.g., seismic fault lines), population centers, and access to the grid should be incorporated into planning.
- Safe operation: All nuclear nations, both developed and developing, must strictly adhere to the principles of the World Association of Nuclear Operators (WANO),[22] as this organization is ideally positioned to be at the forefront of the implementation of new safety measures. However, as of now, WANO has no enforcement mechanisms.

- Emergency preparedness: As demonstrated by the events at Fukushima, when planning for nuclear emergencies, all contingencies must be accounted for, including backup systems; evacuations; command, control, communication; and the role of the military, police, national guard, etc.[23]
- Security measures: Though recent events have shifted the focus of nuclear security to preparing for natural disasters, it cannot be forgotten that a great human threat exists as well. Nuclear facilities will always be a potential target for terrorism and must be prepared accordingly.
- Safe transport of nuclear fuel: As more nations, especially developing ones, embrace nuclear energy, there will be a marked increase in the transport of nuclear fuels worldwide (fuel rods, the mixed oxide fuel known as MOX, spent waste fuel, vitrified waste, etc.). International standards have already been universally adopted here.
- On-site storage: Spent fuel storage on site is an issue for all nuclear nations, including the United States, which currently does not have a permanent waste disposal center. Standards for dry cask and fuel pool storage must be reviewed to reflect the lessons learned from Fukushima.[24]
- Talent pool: Building an adequate talent pool with the education needed for all aspects of the safe application of nuclear technology is an essential step in establishing a successful nuclear program. This need also extends to developed nuclear nations, where much of the talent pool is close to retirement age.[25]
- Local community consensus: Building local consensus, leading to cooperative agreements with local communities, when locating new reactors and nuclear facilities is important in order to facilitate public trust as well as engender an atmosphere of transparency. This also extends to public education on such issues as radiation standards.
- Liability: An adequate liability and compensation system, covering who pays in event of nuclear disaster, needs to be firmly established

before nuclear operations commence in any nation, whether developed or developing. Essentially, the government must step in as the "payer of last resort" due to the scale of potential liability and compensation in the wake of a nuclear accident. The Convention on Supplementary Compensation for Nuclear Damage was put forth to remedy this situation, though not all nuclear nations (including Japan) have signed it.[26]

- Decommissioning: As reactors mature, increased awareness of potential decommissioning problems is required. In this regard, the Japanese may consider making the Fukushima Daiichi plant an international center for studying decommissioning strategy. The experiences of the Department of Energy and its national laboratories can make an important contribution in this regard.
- Nonproliferation safeguards: These must be strengthened, including on-site monitoring (primarily by the IAEA), especially as nations in proliferation-risk regions employ nuclear power on an ever larger scale.

Key Players and Responsibilities

Now that areas of concern in regard to safety have been documented in the previous section, it becomes necessary to determine where responsibility for fulfilling these obligations lies. The global nuclear energy game indeed has many players spread out among governments, companies, and international organizations. What are their individual responsibilities and concerns?

- Governments: They are the most important entity but, as noted, there are variations of regulatory independence, talent pool, and legal issues (including liability). There are also concerns about conflicts of interest, as governments are responsible for regulation but also for aiding in the commercial advancement of their respective nuclear industries.

	Commercial	Research	Nonproliferation	Safety/Security
Governments				
Int Orgs (IAEA, NEA)				
NGOs (WANO, INPO)				
Vendors (AREVA, Toshiba)				
Operators				
Quasi-Official (NSG)				
U.S. Bilateral				
Educational Establishments				

The above table uses the color coding from best (dark blue), to OK (pale blue), worrisome (yellow), and lacking (red).

Figure 9.4 Who Is Responsible?

- International organizations (e.g., IAEA and NEA): International organizations, especially the IAEA, could potentially play the most crucial role in enhancing worldwide nuclear safety and security. The IAEA has legitimacy among all parties (developed nuclear nations, developing, and newcomers) as an independent body.
- Quasi-official organizations and non-governmental organizations (NGOs): These include such multilateral groups as the Nuclear Suppliers Group, which is a coalition of national governments with the aim of preventing proliferation. There are also organizations to ensure worldwide commercial and operational nuclear safety, such as WANO and the Institute of Nuclear Power Operations (INPO).[27]
- State-backed vendors: Nations such as France, Russia, and South Korea have large state-backed or state-owned companies that are aggressively competing for a share of the world commercial nuclear power market. This has large safety, security, and nonproliferation implications, besides the conflicts of interest that can arise. Also, as their influence in commercial nuclear power increases, U.S. influence may wane on safety, security, and nonproliferation. This is a strategic issue and requires a change in thinking about

how U.S. industry and government interact to advance mutual interests.

- Electric utilities and plant operators: The actual electric utilities are large players in whether or not nuclear power expands. They are also pivotal in the implementation and execution of new safety and security initiatives and must be included in any planning; in essence, they are the front line.
- Inter-governmental agreements: Long-established agreements between advanced nuclear powers (such as the U.S.-Japan nuclear accord[28] and the U.S.-Euratom nuclear accord[29]) have facilitated much scientific cooperation and expertise-sharing over the past decades. As new nations push for nuclear power, there are also technology-sharing agreements established or in the works between advanced nuclear nations and advancing newcomers (e.g., the U.S.-India nuclear accord[30] and the U.S.-UAE nuclear accord[31]).
- Educational establishments: As nuclear power expands worldwide, educational establishments will need to be enhanced and capabilities boosted, as not only does the talent pool originate from here but also much of the groundbreaking research. At the university level, it will also be important to educate the future nuclear leaders of the world on the merits and necessity of safety, security, and nonproliferation norms. In the United States, operators have benefited from the expanded program of community colleges for a technical nuclear degree and one can envision a growing market for education and certification online, perhaps coordinated through the IAEA.

Case Examples: Where Nations Stand

Each nation involved with nuclear power has its own regulatory system, responsibilities, and priorities for the future, which can make coming up with a global standard difficult. The wide range of national situations is described below.

- United States: The United States has a strong and independent regulator (NRC) derived from the lessons learned from TMI and the subsequent creation of the industry self-regulating body INPO. It also has a strong talent pool with a high technical ability. It does not engage in reprocessing and has problems with the implementation of permanent waste disposal (e.g., Yucca Mountain).
- France: This nation has a successful program based on standardization and a partially closed fuel cycle (reprocessing). Its government exercises control over its main operators, AREVA and EDF, and is aggressive in pursuing international sales.
- UK: The United Kingdom has a successful program and has plans to expand with a decently sized pool of talent. It engages in reprocessing (for other nations) as well as limited international commercial endeavors.
- Russia: A large and scientifically diverse talent pool remains as a legacy of the Soviet Union. But serious safety, regulatory, and nonproliferation concerns exist as Russia pushes aggressively into international commercial markets in its current incarnation of the ROSATOM state nuclear energy corporation.[32]
- South Korea: This is a quickly growing program (both in size and level of expertise) with a good safety, security, and nonproliferation record (though there are some concerns about material diversion), but it wants to engage in domestic reprocessing. It is also a new but powerful player in international commercial deals (e.g., with the UAE). Korea Hydro and Nuclear Power Co. recently announced its intent to develop a reactor with enhanced safety features.[33]
- Japan: A previously strong proponent of nuclear power, Japan has a diverse and large program with the obvious complications arising out of the Fukushima accident. It will need nuclear power in the long run to meet its energy demands, but the implications of Fukushima and lessons learned are far from settled.
- China: Asia's largest economy has a program that is rapidly expanding (both in size and level of technology) as it races to meet electricity demand. Though it is not a large player in international

commercial markets yet, there is potential.[34] China is working on indigenous designs and intends to be a global supplier.

- India: This nation is a huge potential market but currently the program is quite minimal. There are large nonproliferation concerns on the political side and liability/compensation issues in regard to international vendors successfully beginning projects.[35]
- South Africa: This is a small program, but the government is very keen on expanding via international vendors. Also, South Africa is currently in talks with other governments, including the United States, about inter-governmental agreements.

Integrated Approaches to Global Safety and Security

With this expansion, especially in newcomer nations, certain questions arise. Primarily, how can a nation without the necessary technological, regulatory, and infrastructure development successfully create a safe and secure nuclear program? Nuclear nations would be well advised to follow the guidelines laid out in the 2007 IAEA report, *Considerations to Launch a Nuclear Power Programme.*[36]

As demonstrated, many of the nations and regions where the primary growth in nuclear power is projected to occur have serious infrastructure, regulatory, knowledge, and safety/security shortcomings. It will be essential to address these areas of concern if these programs are to be safe and secure. In these newcomer nations, will there be fuel-take-back arrangements? Will there be local storage and/or a geologic repository for used fuel? Should newcomer nations be allowed to enrich their own fuel? What are the nonproliferation implications? For this nuclear expansion to be successful, especially among newcomer nations, these issues must be resolved and the fuel cycle has to be closed in one way or another (either through reprocessing of spent fuel or permanent disposal).

Safety and security, especially among newcomer nations, will be critical to the advancement of nuclear power globally. The natural place for the coordination of safety and security programs is the IAEA, which already has substantial infrastructure and experience in this regard. But while the IAEA has "watchdog" status for safeguards, the same is not true for safety in general, an important distinction.

It is one thing to identify the specific measures that need to be implemented. It is another to apply them systematically, fairly, and globally. There are, of course, different degrees of balance between national self-interest and governance and international measures that benefit the "law of the commons," basically individual actions by nations that benefit themselves as well as the greater world community. Anybody can come up with a checklist of what needs to be done. The key will be implementation. This could be quite complicated in terms of safe nuclear power operation globally.

Three Possible Scenarios

Two approaches to this situation could be described as "business as usual" and "strengthened international conventions," discussed below. A third possibility is described as "a novel approach."

Business as usual: First, one could consider the current situation as a "business as usual" case. Nuclear power is expanded primarily via commercial operations and priorities under a patchwork of national and international standards. Safety and security measures are not standardized and are a hit-or-miss proposition. This is an incremental approach where commercial interests and bidding set the standard in many countries. The advantage of this is that nuclear power is primarily in the hands of the host nation. Respectable teams can be assembled internationally to assist in development; and training centers can be enhanced, even put online. In this world we would expect that vendors would play an important role in furnishing turnkey operations. The main consideration of the buyer is likely to be price.

A key issue is how maintenance, operation, and safety reviews will be established down the road as the task of running the reactor is left to the host nation. Regulatory systems will be established but their independence and effectiveness will vary. The opportunities for corrupt practices in non-transparent situations may be prevalent, working against the interests of nations such as the United States that have strong foreign anti-corruption legislation. The United States can use its bilateral diplomatic powers in exchanges with new nuclear energy states to encourage safe nuclear development. A key issue will be the extent to which new nuclear states want to have domestic fuel-cycle ability, something that the United States has sought to minimize for over four decades.

Strengthened international conventions: A second approach is to negotiate a stronger international safety and regulatory system with teeth. Clearly the international community is moving in the direction of standardizing safety, but we do not have the legal enforcement frameworks necessary to ensure compliance. International standards with strong legal enforcement by national authorities (the likely mode) or by international organizations (an unlikely mode) can help. Strong penalties may be necessary that can be applied to inadequate construction by vendors or irresponsible operation by operators. This approach would respect the intentions of nations but basically put international nuclear development under international legal guidelines.

We should review the 1994 Convention on Nuclear Safety,[37] the 1986 Convention on Early Notification of a Nuclear Accident,[38] and the 1986 Convention on Assistance in the Case of Nuclear Accident or Radiological Emergency[39] as they might be strengthened in light of the Fukushima accident. The original conventions were established following the Chernobyl accident. Timely review and adjustment in light of Fukushima can strengthen these international standards.

We could also expect under this more aggressive approach that host nations would be encouraged to contract full nuclear services from their vendors. Agreement of vendors to "rules of the road" in the construction and operation of nuclear reactors might benefit the host nation—and under these circumstances we would see an enhancement of WANO. Key to this system would be agreement among key nuclear suppliers to go beyond nuclear supplier group guidelines to a stronger, more comprehensive system that takes a systemic review of a country's nuclear development, including fuel cycle, reactor development, independent regulatory oversight, safe operation, and back-end issues, including waste disposal. This approach is not without political problems. Weapons states have historically not wanted to open up their operations and stockpiles for inspection. Likewise, the United States, France, Russia, and Japan may not wish to abide by new international treaties.

This might become a serious international problem within the IAEA as the delicate balance between safeguards (wanted by OECD) and technology transfer (wanted by developing countries) is shaken, especially if the process is perceived as lecturing developing countries. It also goes back to the old argument similar to climate change and North-South debates: "You industrialized countries had the opportunity to develop your programs as you wished; now you are telling us how to manage our programs." Indeed, this would be the case and there could be issues of pride if outside contractors were given responsibility for cradle-to-grave services. Higher standards would necessarily result in an increase in reactor costs, perhaps adding 5 percent or more to the cost of a plant. The added cost could be split between the host country and the international community.

A novel approach: There might be another option—impractical perhaps but interesting if one continues the multinational approach to an extreme. One could imagine reactors being leased to the purchaser nation with the actual site of the reactor being international ground (like an embassy). The international community would help subsidize the building of the reactor, buy the land, provide adequate safety features, produce electricity, and take responsibility for the backup. The IAEA could have an independent regulatory body (built on NRC principles). The host country would be responsible for purchasing electricity with a long-term contract, enabling the international community to bid out the construction and operation to vendors. The expectation of this international approach would be that the host nation wants electricity and not necessarily the responsibility for running the plant. The international community is reassured that the safety-first principle is applied and that the service would offer "cradle-to-grave" fuel services. The international body governing the construction and operation of the plants can thus begin to systematize and standardize key safety and regulatory concerns and raise the bar for safe operation. The host country would benefit via subsidization from a nuclear energy bank with oversight of the IAEA.

The downside of this approach—and it is a major downside—is that it may reduce the impact of vendors from the United States through greater reliance on large, integrated state vendors, perhaps from Russia, South Korea, France, and, eventually, China. One could argue that a multilateral approach may reduce safety to the least common denominator and work against the interests of the struggling U.S. industry. Others can legitimately argue that the IAEA is not prepared to manage such an enterprise with burdensome governing procedures. Such a plan for international cooperation would need to be crafted with major input and leadership from the U.S. government and industry to ensure that safety, security, and nonproliferation efforts are highly maintained.

These three approaches—and they are not mutually exclusive—illustrate the challenge of balancing national sovereignty and multilateral objectives; considering the economy of electricity production versus the added costs implied by high security, safety, and nonproliferation standards; and integrating the many private sector and government interests to provide for a safer global nuclear energy landscape.

The Role of the United States

The United States was once the leader in the development of nuclear energy technology. However, in the last few decades we have started many things and finished nothing. Each new administration seems to bring in a change in direction. The United States today is not even classified as among those leaders. Westinghouse, whose AP1000 design is the first of the Generation III+ reactors to be licensed in the United States, is a wholly owned subsidiary of Japan's Toshiba, and General Electric works in collaboration with Hitachi on nuclear reactors. We could have a major influence, but not until we get our own house in order and develop a coherent long-range view. An example of a nuclear technology which the United States has "started but not finished" is small modular reactors (SMRs). This highly promising technology (basically small, transportable, contained, and fueled reactors) could become an area where the United States can take the lead.

Where we are still the recognized leader is in the regulation and oversight of our fleet of reactors, including not only the government role, but the role of industry itself. The Three Mile Island accident led to a considerable strengthening of the powers of our NRC and to the creation by nuclear reactor operators of INPO, with a mission "to promote the highest levels of safety and reliability." This includes preparing a workforce for the future, and indeed INPO would be a model for capacity expansion in many countries.

INPO has led to a free exchange of operations information which in turn has enabled the industry to increase reactor "up-time" from about 60 percent before its creation to over 90 percent today. Generation of 50 percent more electricity from the same capital stock is a powerful incentive for cooperation. INPO's example has led to the creation of the similar World Association of Nuclear Operators, though WANO does not have the authority of INPO.

An encouraging sign of continued interest in some sort of U.S. leadership, though mainly in regard to security and nonproliferation rather than energy, was the gathering of world leaders at the request of President Obama in April 2010 to address the issue of accounting for and protecting nuclear materials. Not since the founding of the United Nations had so many leaders gathered on American soil. Fifty leaders pledged to a communiqué with fifty action items. This effort was followed up with a second nuclear summit hosted by South Korea in March 2012. Most of this summit was meant to focus on securing nuclear materials, a wise approach since there are so many conferences on the "lessons of Fukushima" and safety.

At the same time it would be appropriate to introduce the importance of improving the international standards for safety, since all nations will be affected in this increasingly small and interconnected world. For example, the U.S. emergency response teams that can be deployed internationally as well as domestically could serve as the model for a truly international response team, with French, Japanese, and American personnel. Also, in regards to clean-up efforts, the U.S. Department of

Energy has much experience in this area gained from cleaning up former weapons sites. DOE could aid in the current Japanese efforts involving the 800-square-mile zone in Fukushima and also teach its methods to an international response group.

Conclusions

In much the same way that our financial systems are international, so is the nuclear world (energy, weapons, medicine, and advanced technology). The difference between the financial and nuclear worlds is that massive physical damage can result from catastrophic nuclear events, whether they be caused by terrorism, lapses in safety, or natural disaster. These can be intentional, due to human error, or the product of natural events. One recalls the trigger situation of the Cold War, when a small misunderstanding could have led to catastrophic consequences. In a way, we were lucky with the Fukushima accident in that it has been largely contained. This was a remarkable achievement by the Japanese industry and government. But if it can happen in Japan, Ukraine, and the United States, we need to take measures today to reduce the risk, especially as we reduce the broader risk of nuclear weapons and commit ourselves to gradual elimination of all nuclear weapons.

The issues are interconnected and yet the management is fragmented. A systematic approach to these issues needs to be taken. This is not necessarily a technology problem, although nuclear technologies on the horizon with regards to reactors (SMR Gen-IV) and fuel cycle technology will greatly assist our efforts. Rather, the challenge is one of collective responsibility in a global world. Nuclear radiation can cross borders, as was the case with Chernobyl. The consequences of a safety failure cannot necessarily be limited to the place where that failure took place. Thus, safety is not a purely national concern and requires strong international control.

References

1. World Nuclear Association, *Nuclear Century Outlook*, http://www.world-nuclear.org/outlook/clean_energy_need.html.
2. International Atomic Energy Agency, May 2008 report, *The Role of the IAEA to 2020 and Beyond*, http://www.iaea.org/newscenter/news/pdf/2020report0508.pdf.
3. Nuclear Energy Advisory Committee, November 2008 report, *Nuclear Energy: Policies and Technology for the 21st Century*, http://www.ne.doe.gov/neac/neacPDFs/NEAC_Final_Report_Web%20Version.pdf.
4. Nuclear Regulatory Commission's July 2011 Japan task force report, *Recommendations for Enhancing Reactor Safety in the 21st Century: The Near-term Task Force Review of Insights from the Fukushima Dai-ichi Accident*, http://pbadupws.nrc.gov/docs/ML1118/ML111861807.pdf.
5. Energy Information Administration, *Annual Energy Outlook 2011 with Projections to 2035*, April 2011, http://www.eia.gov/forecasts/aeo/pdf/0383%282011%29.pdf.
6. International Energy Agency, World Energy Outlook 2010, http://www.iea.org/weo/2010.asp; Executive Summary, http://www.worldenergyoutlook.org/docs/weo2010/WEO2010_es_english.pdf.
7. International Nuclear Safety Group, *Strengthening the Global Nuclear Safety Regime (INSAG-21)*, 2006, http://www-pub.iaea.org/MTCD/publications/PDF/Pub1277_web.pdf.
8. IAEA, *Fundamental Safety Principles*, 2006, http://www-pub.iaea.org/MTCD/publications/PDF/Pub1273_web.pdf.
9. International Nuclear Safety Group, *A Framework for an Integrated Risk-Informed Decision-Making Process (INSAG-25)*, IAEA, 2011, http://www-pub.iaea.org/MTCD/publications/PDF/Pub1499_web.pdf.
10. International Nuclear Safety Group, *Nuclear Safety Infrastructure for a National Nuclear Power Programme Supported by the IAEA Fundamental Safety Principles (INSAG-22)*, IAEA, 2008, http://www-pub.iaea.org/MTCD/publications/PDF/Pub1350_web.pdf.
11. International Nuclear Safety Group, *Improving the International System for Operating Experience Feedback (INSAG-23)*, IAEA, 2008, http://www-pub.iaea.org/MTCD/publications/PDF/Pub1349_web.pdf.

12. President Dwight D. Eisenhower, "Atoms for Peace" (speech, United Nations General Assembly, December 8, 1953), http://www.iaea.org/About/history_speech.html.
13. U.S. Department of Energy, *Blue Ribbon Commission on America's Nuclear Future: Draft Report to the Secretary of Energy*, July 29, 2011, http://brc.gov/sites/default/files/documents/brc_draft_report_29jul2011_0.pdf.
14. U.S. Department of State, Nuclear Security Summit 2012, http://www.state.gov/nuclearsummit.
15. Institute of Nuclear Power Operations Web site, http://www.inpo.info.
16. Nuclear Suppliers Group Web site, http://www.nuclearsuppliersgroup.org/Leng/default.htm.
17. Nuclear Energy Agency Web site, http://www.oecd-nea.org.

Notes

1. World Association of Nuclear Operators: http://www.wano.info.

2. However, it is interesting to note that the three major nuclear accidents (TMI, Chernobyl, and Fukushima) were all in technologically advanced nations.

3. Judy Dempsey and Jack Ewing, "Germany, in Reversal, Will Close Nuclear Plants by 2022," *New York Times*, May 30, 2011.

4. Source: International Atomic Energy Agency, http://www-pub.iaea.org/MTCD/Publications/PDF/RDS2_web.pdf.

5. Source: IAEA http://www-pub.iaea.org/MTCD/Publications/PDF/RDS2_web.pdf.

6. International Energy Agency 2010 World Energy Outlook: http://www.iea.org/weo/2010.asp. Or see executive summary: http://www.worldenergyoutlook.org/docs/weo2010/WEO2010_es_english.pdf.

7. Jeremy Carl, "Fukushima and the Future of Nuclear Power in China and India," chapter 10, this volume.

8. Alan M. Strout, *The Future of Nuclear Power in the Developing Countries*, MIT Energy Laboratory, 1977, http://dspace.mit.edu/bitstream/handle/1721.1/31246/MIT-EL-77-006WP-04128995.pdf?sequence=1.

9. The economic feasibility of nuclear power is a disputed topic as cost calculations can vary widely depending on externalities (e.g., waste disposal) incorporated into analysis.

10. Carl, "Fukushima and the Future of Nuclear Power in China and India."

11. As noted by the World Nuclear Association, one of the primary reasons the South Korean company KEPCO won the UAE contract over the French company AREVA was cost as well as speed on construction. The report, "Nuclear Power in South Korea," states, "The choice was on the basis of cost and reliability of building schedule. An application for U.S. Design Certification is likely about 2012." See http://world-nuclear.org/info/inf81.html.

12. CoCom stands for the Coordinating Committee for Multilateral Export Controls, which existed following World War II to 1994.

13. Timothy J . Roemer, "France assures NSG waiver for India not undermined," *The Economic Times*, July 1, 2011, http://economictimes.indiatimes.com/news/politics/nation/france-assures-nsg-waiver-for-india-not-undermined/articleshow/9062764.cms.

14. "The Group of 77 (G-77) is the largest intergovernmental organization of developing States in the United Nations. It provides the means for the countries of the South to enhance their joint negotiating capacity on all major international economic issues within the UN and promote South-South cooperation for development. Although the members of the G-77 have increased to 130 countries, the original name was retained because of its historic significance. Chapters of the G-77 have been established in Geneva, Nairobi, Rome, Vienna and Washington, DC," http://www.g77.org.

15. International Atomic Energy Agency, *Reinforcing the Global Nuclear Order for Peace and Prosperity: The Role of the IAEA to 2020 and Beyond*, May, 2008, http://www.iaea.org/newscenter/news/pdf/2020report0508.pdf.

16. http://www-ns.iaea.org/committees/files/insag/743/INSAGLetterReport20117-26-11.pdf.

17. A detailed analysis of the necessary infrastructure is given in an IAEA report, "Milestones in the Development of a National Infrastructure for Nuclear Power," IAEA Nuclear Energy Series NG-G-3.1, 2007. It is quite detailed and few of the nations new to nuclear power meet a fraction of its criteria.

18. Nuclear Regulatory Commission's July 2011 Japan task force report, *Recommendations for Enhancing Reactor Safety in the 21st Century: The Near-term Task Force Review of Insights from the Fukushima Dai-ichi Accident*, http://pbadupws.nrc.gov/docs/ML1118/ML111861807.pdf.

19. Nuclear Energy Advisory Committee, *Nuclear Energy: Policies and Technology for the 21st Century*, November 2008, http://www.ne.doe.gov/neac/neacPDFs/NEAC_Final_Report_Web%20Version.pdf.

20. IAEA *Considerations to Launch a Nuclear Power Programme,* 2007, p. 9, http://www.iaea.org/NuclearPower/Downloads/Launch_NPP/07-11471_Launch_NPP.pdf.

21. NRC, Japan Task Force Report.

22. WANO, "Our Principles," http://www.wano.info/about-us/our-mission.

23. NRC, Japan Task Force Report.

24. Blue Ribbon Commission on America's Nuclear Future, *Draft Report to the Secretary of Energy,* July 29, 2011, http://brc.gov/sites/default/files/documents/brc_draft_report_29jul2011_0.pdf.

25. Nuclear Energy Advisory Committee, *Nuclear Energy Policies and Technology for the 21st Century,* http://www.ne.doe.gov/neac/neacPDFs/NEAC_Final_Report_Web%20Version.pdf.

26. IAEA, Convention on Supplementary Compensation for Nuclear Damage, http://www.iaea.org/Publications/Documents/Infcircs/1998/infcirc567.shtml.

27. http://www.inpo.info.

28. Text of U.S.-Japan nuclear accord available at: http://nnsa.energy.gov/sites/default/files/nnsa/inlinefiles/Japan_123.pdf.

29. Text of the nuclear accord between the United States and the European Atomic Energy Community available at: http://nnsa.energy.gov/sites/default/files/nnsa/inlinefiles/Euratom_123.pdf.

30. Text of U.S.-India nuclear accord available at: http://frwebgate.access.gpo.gov/cgi-bin/getdoc.cgi?dbname=109_cong_bills&docid=f:h5682enr.txt.pdf.

31. Text of U.S.-UAE nuclear accord available at: http://www.fas.org/man/eprint/uae-nuclear.pdf.

32. ROSATOM Web site: http://www.rosatom.ru/en.

33. China Radio International, "S. Korea's Nuke Power Operator Aims for Safer Reactor," August 31, 2011, http://english.cri.cn/6966/2011/08/31/2743s656094.htm.

34. Carl, "Fukushima and the Future of Nuclear Power in China and India."

35. Ibid.

36. IAEA, *Considerations to Launch a Nuclear Power Programme.*

37. http://www-ns.iaea.org/conventions/nuclear-safety.asp.

38. http://www.iaea.org/Publications/Documents/Infcircs/Others/infcirc335.shtml.

39. http://www.iaea.org/Publications/Documents/Conventions/cacnare.html.

Fukushima and the Future of Nuclear Power in China and India

10

JEREMY CARL

Executive Summary

The March 2011 Fukushima disaster is unlikely to fundamentally alter plans by either China or India for a dramatic expansion of nuclear power.

Nuclear power has traditionally been only about 2 percent of each country's energy mix, but it has recently been in rapid ascent. In India, two key events shed light on the probable pace of development. The U.S.-India nuclear deal of 2008, which ended a decades-long embargo and allowed India to access nuclear technology and fuel, has moved the Indian nuclear industry forward substantially from where it was just a few years earlier. The extraordinary lengths that the Indian government went to in order to cement the deal (even risking the government itself in a narrowly won no-confidence vote in parliament) indicate the centrality of nuclear power to India's energy development strategy. Some government sources have forecast nuclear power generation to be 25 percent of India's total power generation by 2050.

India's nuclear power industry has added some new safety procedures in the wake of Fukushima and has opened the curtain to a greater degree on the status and potential risks of its nuclear development strategy. Nonetheless, Fukushima has done little to slow Indian nuclear development. The "first pour" on a new, domestically designed reactor in the Indian state of Rajasthan took place in July 2011, just a few months after the disaster.

At the same time, the failure of Indian policymakers to give acceptable liability coverage to international nuclear companies, despite heavy pressure from the United States at the highest diplomatic levels, indicates the limits of the possible within the Indian political system. Granting liability limitations, always a thorny political issue, became even more difficult after Fukushima. But even more than Fukushima, the politics of Indian nuclear development—in particular the politics of liability waivers—is still haunted by the specter of the Bhopal industrial disaster of 1984. That event killed over 10,000 people and sickened hundreds of thousands of others while permanently souring much of the Indian population on potentially dangerous industrial development led by foreign corporations.

Aside from Fukushima and the legacy of Bhopal, if anything slows nuclear development in India it is likely to be democratic deficits, including rampant corruption and the enduring role of caste politics as a determining factor in Indian elections. These problems are joined by land squabbles, which continue to fell numerous Indian industrial projects. These difficulties can be seen in the setbacks India has had executing on its proposed nuclear energy park in Jaitapur, a 10 gigawatt (GW) agglomeration of six nuclear reactors. If built, it will be the world's largest nuclear power station and a model for several other nuclear energy parks that India plans to construct. The costs of Jaitapur alone are expected to run more than $20 billion.

India's nuclear safety record does not inspire great confidence. With two serious accidents in the 1990s and no full safety audits released to the public, there is substantial reason for concern about India's ability to provide nuclear power safely. While India does possess a skilled workforce of nuclear engineers, it struggles—as many countries do—to keep competent regulators from leaving their positions to work in the nuclear industry, where they can receive far higher salaries and benefits.

China faces many of the same difficulties in the nuclear sphere that plague India. Home to approximately 40 percent of the world's current planned nuclear power plant construction, China' s safety record is largely shrouded in secrecy. Given China's notorious corruption on construction and infrastructure projects—most recently, problems with a high-speed rail scandal that has caused a major political backlash—there is great reason for concern. Even more worrisome, the CEO of China's largest nuclear company was recently jailed for life in a massive bribery scandal involving potential payments and bid-rigging for power plant construction.

China plans to commission a staggering 40 GW of new nuclear power plants by 2015, though this may be slowed to a degree by Fukushima, which caused all nuclear power plant approvals in the country to be frozen. There is substantial concern among both domestic and foreign policymakers about China's lack of trained regulatory personnel, and the difficulty of keeping trained personnel within the regulatory sphere when the financial inducements to join industry can be substantial. In addition, China's primary current state regulator of nuclear power, the National Nuclear Safety Administration, remains underpowered and is situated in the weak environmental ministry.

Even before Fukushima, there seemed to be some moves within China to slow the dramatic nuclear expansion, an expansion that had become so fast that construction had even begun on some plants not formally approved. China's nuclear regulator underestimated the number of power plants under development—plants that were often used as prestige projects by local governments. More than twice the authorized number of projects were under some form of development in 2010, causing China's National Energy Administration (NEA) to raise its five-year plan target to 86 GW before lowering it in the wake of Fukushima. But even with the slowdown, China looks to become the second- or third-largest nuclear power provider in the world within the decade. To achieve its ambitious goals, the government has increasingly moved to involve its top conventional power plant construction companies (the so-called Big Five) in the nuclear build-out, using their ability to access international capital markets to drive growth within the sector.

Due to the closed nature of Chinese governance, far less is known about its accident records than those of India, although any major accident would have been very difficult to hide. There was a serious incident at a nuclear power plant sixty miles south of Shanghai in 1998—an accident that caused the plant to shut down for more than a year. But in general, China's nuclear safety record has not yet suffered any serious public blemishes.

Overall, it is clear that Fukushima has had only a limited impact on China's and India's nuclear power development plans. What impact it has had will probably be a substantial net positive, causing a temporary slowdown in development while the industry re-examines its safety and construction practices from a safety perspective. Yet, absent fundamental changes in the safety culture and institutional incentives in both countries, the chances of a repeat of Fukushima (or worse) on Chinese or Indian soil are not at all insignificant.

Given the potential dangers, both to the lives of millions of Chinese and Indians and, secondarily, to the survival of nuclear power globally, the United States and other interested parties should do everything possible to bolster the safety and regulatory capacities of China's and India's nuclear power industries. Taking aggressive action can help ensure that the possibilities of a Fukushima repeat in the world's two most populous countries are kept to a minimum.

■ ■ ■

China and India are racing on the road toward a nuclear-powered future, and Fukushima is little more than a small but unexpected speed bump—enough to cause political leaders to temporarily ease off of the accelerator, but not significant enough for either country to take a major detour.

While the Fukushima disaster caused each nation's political leaders to take symbolic steps to assure their citizenry of the safety of their nuclear power plants, all public signs indicate that Fukushima is unlikely to seriously alter either country's plans for dramatic expansion of nuclear power. This paper examines the existing Chinese and Indian nuclear fleets and their respective plans for expansion, as well as how these plans were affected by the incident in Japan.

As of 2011, nuclear power was not a large part of the electricity generation mix in either China (11.3 GW) or India (4.8 GW), although this is a reflection more of the enormous scale of each country's overall power systems rather than an indication that either country neglects nuclear power per se. Nonetheless, in contrast to nations like France, whose energy landscape is dominated by the atom, or even the United States, where nuclear energy provides 20 percent of electricity, nuclear energy is much less central to the current energy infrastructures of China and India. It provided approximately 2 percent of both India's and China's power generation as of 2011.[1] While the overall size of each country's electricity markets means that each is still a substantial nuclear player, neither country has ever had nuclear energy at the core of its energy development strategy. But this is changing rapidly as accessing fossil reserves becomes more problematic for both environmental and strategic reasons.

A Brief Overview of Nuclear Power in India

Nuclear energy policy in India is dictated in many respects by two disparate incidents that occurred a quarter of a century apart: the Bhopal disaster of 1984 and the U.S.-India nuclear deal of 2008. And it could easily be argued that Bhopal, which occurred in Madhya Pradesh state in the heartland of India, rather than Fukushima, remains the dominant brake on India's nuclear ambitions.

India currently has 4.8 GW of nuclear power plants in operation, with an additional 3.1 GW under construction—evidence of an aggressive ramp-up rate.[2] India hopes to increase its installed nuclear power capacity to 20 GW by 2020. This would be a substantial increase that would still be a moderate portion of India's electricity generation (about 15 percent of today's total generation), but would do much to diversify India's current coal-based portfolio and strengthen the country's overall energy security.[3] Based on previous performance against plan, a total of 11 GW might be more realistic to expect in this time frame.[4] Some Indian government officials have set goals of generating more than 25 percent of India's total energy from nuclear power by 2050.[5] These plans and targets would have been impossible to consider seriously during the previous Nuclear Suppliers Group (NSG) freeze-out of India, as India's nuclear power stations were chronically fuel-short and ran at far less than full capacity.

In the wake of the U.S.-India nuclear deal in 2008, overall trade velocity increased with new nuclear power deals being signed on a regular basis. Nuclear plant utilization increased from 71 percent to 78 percent in the year following the deal as more reliable fuel supplies became available.

India has developed 540 MW reactors indigenously, and is in the process of developing a fast breeder reactor scheduled to become operational sometime this year. The lack of uranium fuel has long been a constraint on India's nuclear program—access to uranium supplies was one of the primary drivers for the U.S. deal. However, one of the world's largest deposits of uranium was found in southern India in July 2011. This promises to add further momentum to India's nuclear program.

The initial results are now being seen on the ground. In March 2009,[6] India received its first nuclear shipment from foreign suppliers since the decades-long foreign embargo began, obtaining sixty tons of uranium from the French group AREVA, the world's largest builder of nuclear power plants.[7] Two days later a senior French naval officer endorsed India's development of a nuclear-powered submarine attack fleet. A week before the AREVA deal, BHEL and Nuclear Power Corporation of India, both companies run by the Indian government, signed agreements with GE Hitachi agreeing to purchase nuclear reactors from it as well.

The Legacy of Bhopal

All major industrial developments in India exist in the historical shadow of the Bhopal disaster which, while not a nuclear accident, was perhaps the worst conventional industrial accident of modern times. In 1984, a chemical plant run by U.S.-based multinational corporation Union Carbide accidentally released poisonous methyl isocyanate over the unsuspecting central Indian city of Bhopal, ultimately killing more than 10,000 people and injuring hundreds of thousands more.[8] While assignment of blame for the accident has gone back and forth over the years, the incident caused a long-standing public mistrust in India of large industrial projects that hold potential health risks—especially when these projects, like many nuclear power plants, are run by multinationals. Furthermore, the total damages paid by Union Carbide were considered woefully inadequate by both foreign and domestic observers.[9] (Damages of just $470 million in 1989 dollars—approximately $859 million in 2011 dollars—were ordered paid by the Supreme Court of India in exchange for all claims.) All of this contributed to a feeling among many Indians not only of fear of an industrial accident but a lack of confidence that the government would see to it that they or their families were appropriately compensated if a disaster were to happen.

Another concern for both Indian citizens and international policymakers is a pervasive culture of bribery, dangerous shortcuts, and fraud in Indian construction and other industrial projects and a casualness about the release of pollution that many view as incompatible with running a safe and advanced nuclear program. In addition, regulatory capture, while a problem worldwide, has been particularly acute in India, with notorious recent examples in both the mining and wireless communications sectors.

The U.S.-India Nuclear Deal

While Bhopal casts a shadow over India's industrial order, the completion of the U.S.-India nuclear deal is an indication of the value India places on nuclear energy for its strategic energy future. The deal, which brought India out of nuclear exile by creating an exception to the Non-Proliferation Treaty (NPT) that allowed it to access nuclear fuel and technology, was finally approved by the Indian Parliament and the U.S. Congress in August 2008. At the time it was passed in Congress, during the heart of the financial crisis when virtually no other legislation was moving, analysts and media described the successful push for passage as the most active the India lobby had ever been on an issue involving the United States.

Indian Prime Minister Manmohan Singh described the deal as "a cornerstone of the new strategic partnership between the two countries."[10] It faced bitter opposition, largely from the Communist-dominated left-front at home. The internal deal came close to toppling Singh's government during a parliamentary vote of no-confidence in a sordid affair with credible accusations of legislative bribery that continue to reverberate in Indian politics to this day.[11] In ratifying the deal and making an exception to the NPT that would allow India to receive nuclear fuel and technology, the U.S. Congress signed off on a deal widely perceived as being favorable to India.[12]

With numerous foreign vendors lining up eagerly to take part in a nuclear power renaissance in India, and with a large and competent domestic nuclear engineering capability already present, India has the potential for a rapid build-out of its nuclear capability. Ironically, the United States itself has been poorly positioned to take advantage of this opportunity. India's current nuclear liability laws strongly favor state-owned firms that can take on unlimited liability, while not providing sufficient protection for private players from the United States. Secretary of State Hillary Clinton has made the reform of such laws a major cornerstone of the United States' engagement with India, but has obtained little traction.[13]

Indian Nuclear Power After Fukushima

Generally, in the wake of Fukushima, India has continued to aggressively pursue nuclear energy possibilities. Opposition to India's nuclear ambitions on the domestic front has centered on domestic politics more than international concerns. A post-Fukushima committee convened by the government to address the safety and security of India's nuclear power plants announced plans that within nine months all nuclear plants in India will be able to shut down automatically in the event of an earthquake.[14] (It is important to note that these sorts of projects in India are notoriously behind schedule.) A few other basic safety measures were announced. But fundamentally, Fukushima changed little. India was not deterred from its nuclear ambitions, as indicated by the first pour on

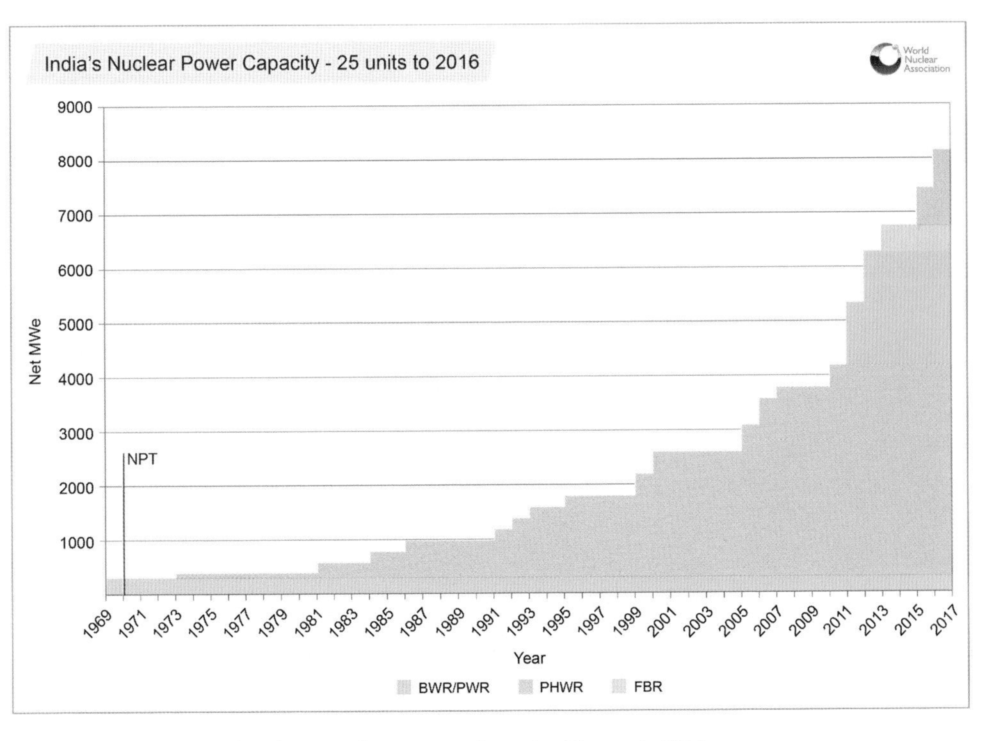

Figure 10.1 Projected Indian Nuclear Power Capacity Through 2016.
Source: World Nuclear Association.

construction of a new nuclear power plant in the state of Rajasthan on July 18, 2011.[15]

The major impediment to future development of nuclear power in India remains democratic deficits and land squabbles. Indian politicians are often poorly responsive to their constituents, and land acquisition difficulties have plagued almost all major Indian industrial projects as of late. Much of this dysfunctionality is a product of India's dense population and impossibly arcane land laws, many of which date from the colonial period.[16]

These problems can be seen in the difficulties surrounding the planned massive 10 GW nuclear plant at Jaitapur, scheduled to be built at a cost of over $20 billion in the industrial Maharashtra state, home to Mumbai, India's financial capital. The Jaitapur project, which was supposed to be the start of India's nuclear renaissance, has instead become the locus of sometimes violent protests against nuclear expansion. The protests are led by a combination of local activists and representatives from the far-left Communists and the communally oriented Hindu radical group Shiv Sena.[17]As environmentalists argue with bureaucrats, villagers demand more compensation, and plant equipment makers demand greater liability protection, it seems unlikely that the project will begin on time.

India's nuclear establishment sees Jaitapur as the template for a series of five or more nuclear energy parks, each with a capacity for up to 10 GW that would make any of these nuclear energy parks the largest in the world. [18] Given the difficulties of siting and executing on Indian nuclear projects, concentrating development in a few areas may be a sensible strategy, although Fukushima also shows the risks of such an approach if one reactor fails.

If it is constructed, the 10 GW Jaitapur plant complex, to be built by AREVA, would become by far the world's largest. Jaitapur will involve the displacement of more than 10,000 people from almost 1,000 hectares of land that will be necessary for the project.[19] This one project alone in the rural district would represent almost 5 percent[20] of India's installed base of power and likely considerably more of its actual generation given nuclear power's high availability. "Nuclear energy is the cheapest and cleanest way to get India's cities on the grid system," said Jairam Ramesh, who until recently was India's star environmental minister and previously had served as minister of state for power.[21] Ramesh has been an aggressive booster of nuclear power in India.

India hopes to increase nuclear's share to 25 percent by 2050, though most independent analysts feel these plans are unrealistic.[22] Liability laws passed by the Indian parliament in the wake of the deal have deterred U.S.-based companies from tapping into the potentially lucrative market. These laws potentially place equipment suppliers at almost unlimited legal risk in the event of an accident. As one commentator wrote, "The UPA (United Progressive Alliance) government can't contemplate diluting the liability laws due to massive domestic pressure. Virtually the entire opposition, from left to right, has pressured Prime Minister Manmohan Singh into hardening liability legislation. Tinkering with this now would therefore be political suicide for his government."[23] The nuclear liability issue in India has become embroiled in broader strategic matters involving India's assertions of its sovereignty and governmental survival.

History of Nuclear Accidents in India

India has had several nuclear accidents in its history, most notably a serious 1993 incident at the Narora plant in which a fire broke out after a blade broke off a high-speed steam turbine. As with Fukushima, a blackout and near-meltdown occurred, although with less catastrophic consequences. Another serious incident at India's nuclear plant in Kota, Rajasthan, shut the plant down for two years in 1995.

Past Nuclear Safety Audits in India

Prior to Fukushima, India had three major safety audits of its nuclear power operations, two directly linked to major nuclear incidents in other countries. However, in a reflection of India's relatively closed political culture, none of those major audits have been made public, an unfortunate situation that India looks to change with the post-Fukushima audit. The first audit was conducted in the wake of the Three Mile Island incident

Table 10.1 Notable Nuclear Power Plant Accidents in India[24]

Date	Location	Description	Cost (in millions 2006 U.S.$)
4 May 1987	Kalpakkam, Tamil Nadu, India	Fast Breeder Test Reactor at Kalpakkam refueling accident that ruptures the reactor core, resulting in a two-year shutdown.	300
10 September 1989	Tarapur, Maharashtra, India	Operators at the Tarapur Atomic Power Station find that the reactor had been leaking radioactive iodine at more than 700 times normal levels. Repairs to the reactor take more than a year.	78
13 May 1992	Tarapur, Maharashtra, India	A malfunctioning tube causes the Tarapur Atomic Power Station to release 12 curies of radioactivity.	2
31 March 1993	Bulandshahr, Uttar Pradesh, India	The Narora Atomic Power Station suffers a fire at two of its steam turbine blades, no damage to the reactor. All major cables burnt.	220
2 February 1995	Kota, Rajasthan, India	The Rajasthan Atomic Power Station leaks radioactive helium and heavy water into the Rana Pratap Sagar dam, necessitating a two-year shutdown for repairs.	280
22 October 2002	Kalpakkam, Tamil Nadu, India	Almost 100 kg radioactive sodium at a fast breeder reactor leaks into a purification cabin, ruining a number of valves and operating systems.	30

in 1979. The results of that probe were not made public, indicative of India's more closed politics during that time. In the wake of Chernobyl in 1986, a similar audit was done, and again largely shelved as "top secret." In 1995, the first public audit of India's nuclear facilities was undertaken at the instigation of Dr. A. Gopalakrishnan, then chairman of India's Atomic Energy Regulatory Board. Despite pressure by him to release it, this was again labeled "top secret." Since Fukushima, the chairman has called for independent IAEA assessments of India's nuclear capability rather than ones "conducted solely by a captive and relatively inexperienced AERB or through a peer review by the World Association of Nuclear Operators, a fraternal partner of NPCIL whose past opinions are mostly found to be biased in favour of NPCIL."[25] The public release of early recommendations post-Fukushima may be the welcome sign of greater openness in India's nuclear establishment.

Overview of China's Nuclear Industry

Given the lack of democratic accountability in China, it is unsurprising that China's official reaction to Fukushima was even more muted than India's, though notably the government did respond at least somewhat to public pressure to slow the scale of nuclear development. But even a slowdown will not alter the fundamentals: China has 40 percent of the world's currently planned nuclear power projects under construction.[26]

Furthermore, China's leadership has said it will not substantially alter its plans to build new reactors in the wake of Fukushima. The recently released five-year plan calls for 40 GW of new nuclear power by 2015.[27]

But even if Fukushima will not change China's strategy in the nuclear arena, the event underlines its risks. As Bo Kong and David Lampton, two American scholars who have recently completed an intensive study of China's in-progress nuclear plant development, observed, "Beijing must rein-in its runaway nuclear industry, rebalancing speed and safety."[28]

This was brought home by the recent imprisonment and stripping of party membership of Kang Rixin, the president of China National Nuclear Corporation (CNNC) and a member of the Party's central committee, who was implicated in a massive $260 million bribery scandal involving bid-rigging for power plants.[29]

China's existing nuclear power installed capacity is roughly on par with India's, which is surprising given that China's overall power grid is five times as large as India's.[30] However, given the pace of new construction, China seems likely to soon join the United States and France among the nuclear giants.

As of 2011, China had thirteen nuclear power plants at four sites, with twenty-seven more in various stages of construction.[31] The National Development and Reform Commission (NDRC), China's most powerful planning body, has set a goal of raising nuclear power to 6 percent of Chinese electricity output by 2020. If this expansion were to take place, it would make China the largest supplier of nuclear power in the world after the United States.[32] Most of China's new reactors are CPR-1000s, with fifteen units under construction as of June 2010. These are closely derived from AREVA's 900 MW reactors.[33]

History suggests, however, that even before Fukushima, China's aggressive expansion plans might not prove realistic, even for a country with China's enormous ambitions and skill at rapid-fire infrastructure construction. In addition to the capacity of domestic firms and international suppliers to provide complex nuclear equipment at this scale and velocity, China itself would have a hard time procuring the necessary fuel, equipment, and trained maintenance and safety workforce to increase its plant and equipment at that scale.

Meanwhile, China has less than one-tenth of the nuclear regulatory workforce that is present in the United States and only a small fraction of the budget of the Nuclear Regulatory Commission that regulates the industry in the United States. China has budgeted for another 700 regulatory staff, but properly training such staff inevitably takes time. Expansion of the workforce has been slow; China often loses its top regulators to companies that can afford to pay higher salaries. Furthermore, the location of China's key nuclear regulatory body in the Ministry of Environmental Protection ensures that it will be bureaucratically weak, a fault that China is attempting to address through current reforms. In addition to its general regulatory weakness, China lacks an overall nuclear law and regulatory responsibility is diffused among ten agencies.[34]

To meet capital needs for construction, China has begun granting licenses for nuclear projects to China's Big Five power corporations: Huaneng, Huadian, Datang, Guodian, and China Power Investment Group. Like the two existing nuclear companies—China National Nuclear Corporation (CNNC) and China Guangdong Nuclear Power Group (CGNPG)—the Big Five are state-owned enterprises (SOEs). But unlike

China's pure-play nuclear companies, the Big Five have listed subsidiaries in Hong Kong, giving them better access to international capital, necessary for undertaking massive nuclear power projects.[35]

In response to out-of-control construction, China's National Energy Agency (NEA) was forced to raise the 2020 target for nuclear build-out to 86 GW, only to find that an estimated 226 GW of new capacity was in various stages of planning. Some plants—including Rushan (Shandong Province) and Jiujiang (Jianxi Province)—had begun preliminary work without even obtaining approval from China's National Nuclear Safety Agency (NNSA).[36] Such legal evasions are common in Chinese power construction, as was witnessed by notorious unofficial coal plants in Inner Mongolia and elsewhere in the last decade.[37] However, it's one thing to play fast and loose by permitting thermal power plant construction; doing the same with nuclear plant construction is a much more serious matter.

Nuclear Accidents and Safety in China

There are inherent safety problems in the chosen locations of China's nuclear power plants. Several are near earthquake faults. China is notoriously quake-prone, although its quakes have not historically had the severity of the 9.0 quake that decimated Fukushima. Nonetheless, there is intense debate within China as to whether its nuclear power plant fleet—both currently built and under construction—is sufficiently engineered to withstand the earthquakes it is likely to encounter. He Zuoxiu, a member of the Chinese Academy of Sciences and one of China's most prominent physicists, has publicly and strongly criticized safety standards in Chinese nuclear power plant construction. In particular, he has taken aim at earthquake safety, stating that power plants have not been designed with sufficient earthquake safety in mind because of the substantial increase in capital costs this would entail.[38] Publicly, Chinese officials claim that China's new Generation III reactors with passive cooling will be immune to the problems that plagued Fukushima after the earthquake and tsunami. But comments from leading figures such as Dr. He create concern that China's nuclear generals may be fighting the last war.

There are no *publicly known* major nuclear accidents in China's history. One moderately serious accident that was made public was a 1999 accident at a Shanghai nuclear power plant that left that plant non-operational for a full year.[39] An accident involving improper welding caused some pipes to fall off the reactor under strong water pressure. While Chinese officials insisted that the accident did not rise to the "level 2" requirement at which point it would be mandatory to share with the IAEA, neighboring governments, especially Japan, were highly critical of Chinese secrecy about the accident—and of the late timing of its disclosure. China does have informal channels through which it shares nuclear information with some regional governments, but this is always done on the condition that this information not be made public.[40] While in theory this is reassuring, it is important to note that China tightly controls the state media, and only the most serious accident would likely leak publicly. Nonetheless, it is notable that China has generally invested significantly more in safety in prestigious and high-visibility areas than it has in general industrial projects. For example, China's recent safety record in commercial aviation has been excellent.[41]

China's Reaction to Fukushima

China's reaction to Fukushima was relatively muted. The State Council announced shortly after the scale of the Fukushima disaster became clear that China would suspend approval for all new nuclear power plants while conducting safety checks on all existing plants both in operation and, more importantly, under construction.[42] However, few observers believe that Fukushima will bring anything more than a momentary pause in China's nuclear build-out.

Xie Zhenua, vice chairman of the NDRC, claimed that there would be no slowdown of nuclear construction; Premier Wen Jiabao said that medium- and long-term nuclear plans would be "adjusted and improved."[43]

China is very likely to resume construction of nuclear plants this year.[44] Feng Yi, deputy secretary-general of the China Nuclear Energy Association, stressed that China's nuclear development would slow for

Table 10.2 Chinese Nuclear Power Plants Under Construction

Plant	Province	MWe Gross	Reactor Model	Project Control	Construction Start	Operation
Qinshan Phase II unit 4	Zhejiang	650	CNP-600	CNNC	**1/07**	2012
Hongyanhe units 1–4	Liaoning	4x1080	CPR-1000	CGNPC	**8/07**, **4/08**, **3/09**, **8/09**	10/12, 2013, 2014
Ningde units 1–4	Fujian	4x1080	CPR-1000	CGNPC, with Datang	**2/08**, **11/08**, **1/10**, **9/10**	12/12, 2013, 2014, 2015
Fuqing units 1&2	Fujian	2x1080	CPR-1000	CNNC	**11/08**, **6/09**	10/13, 8/14
Yangjiang units 1–4	Guangdong	4x1080	CPR-1000	CGNPC	**12/08**, **8/09**, **11/10**, 15/3/11	8/13, 2014, 2015, 2016
Fangjiashan units 1&2	Zhejiang	2x1080	CPR-1000	CNNC	**12/08**, **7/09**	12/13, 10/14
Sanmen units 1&2	Zhejiang	2x1250	AP1000	CNNC	**3/09**, **12/09**	11/13, 9/14
Haiyang units 1&2	Shandong	2x1250	AP1000	CPI	**9/09**, **6/10**	5/14, 3/15
Taishan units 1&2	Guangdong	2x1770	EPR	CGNPC	**10/09**, **4/10**	10/13, 11/14
Fangchenggang units 1&2	Guangxi	2x1080	CPR-1000	CGNPC	**7/10**, 2011	2015, 2016
Fuqing units 3&4	Fujian	2x1080	CPR-1000	CNNC	**7/10**, 2011	7/15, 5/16
Fuqing units 5&6	Fujian	2x1080	CPR-1000 or CNP1000	CNNC	?, ?	-

(continued)

Table 10.2 Chinese Nuclear Power Plants Under Construction (*continued*)

Plant	Province	MWe Gross	Reactor Model	Project Control	Construction Start	Operation
Changjiang units 1&2	Hainan	2x650	CNP-600	CNNC & Huaneng	**4/10, 11/10**	2014, 2015
Hongshiding (Rushan) units 1&2	Shandong	2x1080	CPR-1000	CNEC/CNNC	Deferred from 2009?	2015
Yangjiang units 5&6	Guangdong	2x1080	CPR-1000	CGNPC	2011?	2017
Ningde units 5&6	Fujian	2x1080	CPR-1000	CGNPC		
Xianning (Dafan) units 1&2	Hubei	2x1250	AP1000	CGNPC	2011 or 2015	2015?
Taohuajiang units 1–4	Hunan	4x1250	AP1000	CNNC	2011	4/2015–2018?
Pengze units 1&2	Jiangxi	2x1250	AP1000	CPI	2011 or 2015	2015?

two to three years but that its medium- and long-term strategy would be unaltered by Fukushima.[45]

Nonetheless, even before Fukushima there were some signs of unease about the pace of nuclear build-out from China's policy establishment. In January 2011, the State Council Research Office (SCRO), which guides the state council on policy research, urged a slowdown to a 2020 target of 70 GW from the 86 GW currently planned because of concerns about quality control.[46]

Conclusion

Both China and India seem to be proceeding apace with their plans for nuclear power even after the Fukushima incident. Each has paused to some degree to take a much-needed safety assessment, but any changes are far more likely to be incremental than transformational. Given the poor culture of industrial safety in each country and the rapid pace of nuclear power plant development, the international community should be very concerned about the pace and quality of Chinese and Indian nuclear plans. These concerns relate primarily to safety, though with China there are also legitimate proliferation issues, as China was involved in proliferation controversies with Pakistan, Saudi Arabia, Iran, and Iraq.[47]

Overall, there is little in China's or India's record of industrial safety to inspire great confidence that a broad and rapid expansion of nuclear capability can take place without potentially dangerous safety compromises. If failure can happen in Japan, catastrophe can happen in China and India.

Given the potential advantages of nuclear power with respect to climate change, and the fact that we hold little leverage over China's and India's ultimate nuclear plans, the best possible U.S. path for engagement would be to attempt to do everything possible to bolster the safety and regulatory cultures of Chinese and Indian power stations so that the possibilities of a Fukushima repeat in the world's two most populous countries are kept to a minimum.

References

Basu, Mihika. "N-plants: Safety measures likely in 14 months." *Indian Express,* July 22, 2011. http://www.indianexpress.com/news/nplants-safety-measures-likely-in-14-months/820702.

BBC News. "Chinese Nuclear Accident Revealed." July 5, 1999. http://news.bbc.co.uk/2/hi/asia-pacific/386285.stm.

BP Global. *Statistical Review of World Energy.* 2011. http://www.bp.com/assets/bp_internet/globalbp/globalbp_uk_english/reports_and_publications/statistical_energy_review_2011/STAGING/local_assets/pdf/statistical_review_of_world_energy_full_report_2011.pdf.

Bradsher, Keith. "Nuclear Power Expansion In China Stirs Concerns." *New York Times,* December 15, 2009. http://www.nytimes.com/2009/12/16/business/global/16chinanuke.html?_r=1&partner=rss&emc=rss.

Byerly, Rebecca. "India Maps Out a Nuclear Power Future Amid Opposition." *National Geographic News,* July 22, 2011. http://news.nationalgeographic.com/news/energy/2011/07/110722-india-nuclear-jaitapur.

Central Electricity Authority, Government of India, "All India Regionwise Generating Installed Capacity (MW) of Power Utilities Including Allocated Shares in Joint and Central Sector Utilities." July 2011. http://www.cea.nic.in/reports/monthly/executive_rep/jul11/8.pdf.

Chaudhury, Dipanjan Roy. "Safety Must be Hallmark of Nuclear Power Plan." *India Today,* May 23, 2011. http://indiatoday.intoday.in/site/story/safety-must-be-hallmark-of-nuclear-power-plan/1/138982.html.

Doherty, Ben. "Japanese fallout hits India's plans." *The Age,* April 23, 2011. http://www.theage.com.au/world/japanese-fallout-hits-indias-plans-20110422-1dri3.html.

Ewing, Kent. "Plane Crash Revives China Safety Fears." *Asia Times Online,* September 15, 2010. http://www.atimes.com/atimes/China/LI15Ad01.html.

Gopalakrishnan, A. "Nuclear power: The Missing Safety Audits." *Daily News & Analysis,* April 26, 2011. http://www.dnaindia.com/mumbai/report_nuclear-power-the-missing-safety-audits_1536223-2.

Hazarika, Sanjoy. "Bhopal Payments by Union Carbide Set at $470 Million." *New York Times,* February 15, 1989. http://www.nytimes.com/1989/02/15/business/bhopal-payments-by-union-carbide-set-at-470-million.html.

Hook, Leslie. "China's Nuclear Freeze to Last to 2012." *Financial Times,* April 12, 2011. http://www.ft.com/intl/cms/s/68694fe0-6525-11e0-b150-00144feab

49a,Authorised=false.html?_i_location=http%3A%2F%2Fwww.ft.com%2Fcms%2Fs%2F0%2F68694fe0-6525-11e0-b150-00144feab49a.html&_i_referer=#axzz1V4thCeh7.

Kaldaya.net. "China Unable to Guarantee the Safety of its Nuclear Plants." May 30, 2011. http://www.kaldaya.net/2011/News/05/May30_E5_ChinaNews.html.

Kong, Bo, and David M. Lampton, "How Safely Will China Go Nuclear?" *China-U.S. Focus,* April 6, 2011. http://www.chinausfocus.com/energy-environment/how-safely-will-china-go-nuclear.

Krishnaswami, Sridhar. "U.S. House Votes for Nuclear Deal." *The Hindu*, July 28, 2006. http://www.hindu.com/2006/07/28/stories/2006072812290100.htm.

MSNBC News Service. "7 guilty in Bhopal tragedy that killed 15,000." June 7, 2010. http://www.msnbc.msn.com/id/37551856/ns/world_news-south_and_central_asia/t/guilty-bhopal-tragedy-killed/#.TkRHYIJvaSo.

Nautiyal, Annpurna. "The Indo-U.S. Nuclear Deal: What is There?" *Strategic Insights*, vol. 7, no. 4 (September, 2008). http://www.nps.edu/Academics/centers/ccc/publications/OnlineJournal/2008/Sep/nautiyalSep08.html.

Nuclear Age Peace Foundation. "China." Nuclearfiles.org, http://nuclearfiles.org/menu/key-issues/nuclear-weapons/issues/proliferation/china/index.htm.

Nuclear Control Institute. "China's *Non*-Proliferation Words vs. China's Nuclear Proliferation Deeds." http://www.nci.org/i/ib12997.htm.

Nuclear Power Daily. "India reveals 'world's biggest' uranium discovery." July 19, 2011. http://www.nuclearpowerdaily.com/reports/India_reveals_worlds_biggest_uranium_discovery_999.html.

Oster, Shai. "Illegal Power Plants, Coal Mines in China Pose Challenge for Beijing." *Wall Street Journal*, December 27, 2006. http://www.wsj-asia.com/pdf/WSJA_2007_Pulitzer_China.pdf.

Press Trust of India, "NFC to Receive 60 Tonnes Uranium from AREVA." Outlookindia.com, February 18, 2009. http://news.outlookindia.com/item.aspx?654255.

Press Trust of India. "India's 20th Nuclear Plant Goes Critical." *IBN Live*, November 27, 2010. http://ibnlive.in.com/news/indias-20th-nuclear-power-plant-goes-critical/135995-3.html?from=rhs.

Press Trust of India. "Post Fukushima: India begins construction on new nuclear plant." *The Times of India*, July 18, 2011. http://articles.timesofindia.indiatimes.com/2011-07-18/india/29786349_1_nuclear-plant-emergency-core-cooling-system-mw-phwrs.

Project of the Nuclear Age Peace Foundation. "China." *Nuclearfiles.org*. http://nuclearfiles.org/menu/key-issues/nuclear-weapons/issues/proliferation/china/index.htm.

Reuters. "China suspends approval for nuclear power plants." *The Guardian*, March 16, 2011. http://www.guardian.co.uk/world/2011/mar/16/china-suspends-approval-nuclear-plants.

Sharma, Rajeev. "Clinton's India Nuclear Landmine." *The Diplomat*, July 21, 2011. http://the-diplomat.com/indian-decade/2011/07/21/clintons-indian-nuclear-landmine.

Sinomach, "Installed Electric Power Capacity in China Breaks Through 900 GW." Vol. 13, no. 3 (October 18, 2010). http://www.sinomach.com.cn/templates/T_news_en/content.aspx?nodeid=579&page=ContentPage&contentid=4947.

South Asia Monitor. "India at 60: The India-U.S. Nuclear Deal on Hold—Crash, or Course Correction?" October 26, 2007. http://csis.org/files/media/csis/pubs/sam112.pdf.

Sovacool, Benjamin K. "A Critical Evaluation of Nuclear Power and Renewable Electricity in Asia." *Journal of Contemporary Asia*, vol. 40, no. 3 (August 2010), pp. 393–400.

Sovacool, Benjamin K. "The Accidental Century—Prominent Energy Accidents in the Last 100 Years." *Exploration & Production: Oil & Gas Review*, vol. 7, no. 2 (2009).

Stanway, David. "In China the Big Nuclear Question Is 'How Soon?'" *Reuters*, May 8, 2011. http://solveclimatenews.com/news/20110503/china-nuclear-energy-research-fusion-reactors?page=3.

Suga, Masumi, and Shunichi Ozasa. "China to Build More Nuclear Plants, Japan Steel Says." *Bloomberg News*, September 7, 2009. http://www.bloomberg.com/apps/news?pid=newsarchive&sid=a2lUkzmYNGWI.

The Maverick Blog. "India's Nuclear Polity: Shouldn't We Learn from Fukushima?" March 23, 2011. http://shaktimohapatra.wordpress.com/2011/03/23/indias-nuclear-polity-shouldnt-we-learn-from-fukushima.

Thomas, Cherian, and Bibhudatta Pradhan. "India's Singh Wins Parliament Vote on Nuclear Deal." *Bloomberg News*, July 22, 2008. http://www.bloomberg.com/apps/news?pid=20601087&sid=aHJJfKjcKSvY&refer=home.

Want China Times. "Wenzhou oversights reveal risks of China's nuclear program." August 3, 2011. http://www.wantchinatimes.com/news-subclass-cnt.aspx?id=20110803000007&cid=110.

World Nuclear Association. "Nuclear Power in China." Updated March 2012. Accessed August 12, 2011. http://www.world-nuclear.org/info/inf63.html.

World Nuclear Association. "Nuclear Power in India." Updated February 2012. Accessed August 12, 2011. http://www.world-nuclear.org/info/inf53.html.

World Nuclear Association. "World Nuclear Power Reactors & Uranium Requirements." Updated March 9, 2012. Accessed August 8, 2011. http://www.world-nuclear.org/info/reactors.html.

Xinhua. "China not to change plan for nuclear power projects: government." March 12, 2011. http://news.xinhuanet.com/english2010/china/2011-03/12/c_13774519.htm.

Notes

1. BP Global, *Statistical Review of World Energy*, 2011.
2. Press Trust of India, "India's 20th Nuclear Plant Goes Critical," *IBN Live*, November 27, 2010.
3. India's current energy mix is dominated by coal, which is abundant domestically but difficult to extract in sufficient volumes to meet demand due to various political and environmental factors.
4. Author's calculation is based on previous performance of Indian power plant construction versus five-year plans.
5. World Nuclear Association, "Nuclear Power in India."
6. Press Trust of India, "NFC to Receive 60 Tonnes Uranium from AREVA," Outlookindia.com, February 18, 2009.
7. Ibid.
8. MSNBC News Service, "7 guilty in Bhopal tragedy that killed 15,000," June 7, 2010.
9. Sanjoy Hazarika, "Bhopal Payments by Union Carbide set at $470 Million," *New York Times*, February 15, 1989.
10. Sridhar Krishnaswami, "U.S. House Votes for Nuclear Deal," *The Hindu*, July 28, 2006.
11. Cherian Thomas and Bibhudatta Pradhan, "India's Singh Wins Parliament Vote on Nuclear Deal," *Bloomberg News*, July 22, 2008.
12. See, for example, *South Asia Monitor*, "India at 60: The India-U.S. Nuclear Deal on Hold—Crash, or Course Correction?" October 26, 2007, and Annpurna Nautiyal, "The Indo-U.S. Nuclear Deal: What is There?" *Strategic Insights*, vol. 7, no. 4 (September 2008).
13. Rebecca Byerly, "India Maps Out a Nuclear Power Future Amid Opposition," *National Geographic*, July 22, 2011.

14. Mihika Basu, "N-plants: Safety measures likely in 14 months," *Indian Express*, July 22, 2011.

15. Press Trust of India, "Post Fukushima: India begins construction on new nuclear plant," *The Times of India*, July 18, 2011.

16. For more extensive discussion of this issue, see Rachel Brule, "Land Rights without Laws: Understanding Property Rights Institutions, Growth, and Development in Rural India," Center on Democracy, Development, and the Rule of Law, Freeman Spogli Institute for International Studies, Stanford University, February 2009, http://iis-db.stanford.edu/pubs/22415/No_98_Brule_Land rightswithoutlaws.pdf.

17. Dipanjan Roy Chaudhury, "Safety Must be Hallmark of Nuclear Power Plan," *India Today*, May 23, 2011.

18. World Nuclear Association, "Nuclear Power in India."

19. Ben Doherty, "Japanese fallout hits India's plans," *The Age*, April 23, 2011.

20. Central Electricity Authority, Government of India, "All India Region-wise Generating Installed Capacity (MW) of Power Utilities Including Allocated Shares in Joint and Central Sector Utilities," July 2011.

21. Byerly, "India Maps Out a Nuclear Power Future."

22. *Nuclear Power Daily*, "India reveals 'world's biggest' uranium discovery," July 19, 2011.

23. Rajeev Sharma, "Clinton's India Nuclear Landmine," *The Diplomat*, July 21, 2011.

24. Benjamin K. Sovacool, "A Critical Evaluation of Nuclear Power and Renewable Electricity in Asia," *Journal of Contemporary Asia*, vol. 40, no. 3 (August 2010), pp. 393–400; and Sovacool, "The Accidental Century—Prominent Energy Accidents in the Last 100 Years," *Exploration & Production: Oil & Gas Review*, vol. 7, no. 2 (2009).

25. A. Gopalakrishnan, "Nuclear Power: The Missing Safety Audits," *Daily News & Analysis*, April 26, 2011.

26. Reuters, "China suspends approval for nuclear power plants," *The Guardian*, March 16, 2011.

27. *Xinhua*, "China not to change plan for nuclear power projects: government," March 12, 2011.

28. Bo Kong and David M. Lampton, "How Safely Will China Go Nuclear?" *China-U.S. Focus*, April 6, 2011.

29. Keith Bradsher, "Nuclear Power Expansion In China Stirs Concerns," *New York Times*, December 15, 2009.

30. India has about 180 GW of installed power, while China has more than 900 GW. See *Sinomach*, "Installed Electric Power Capacity in China Breaks Through 900 GW," vol. 13, no. 3 (October 18, 2010).

31. World Nuclear Association, "Power Reactors & Uranium Requirements."

32. Masumi Suga and Shunichi Ozasa, "China to Build More Nuclear Plants, Japan Steel Says," *Bloomberg News*, September 7, 2009.

33. World Nuclear Association, "Nuclear Power in China."

34. Ibid.

35. Ibid.

36. Kong and Lampton, "How Safely Will China Go Nuclear?"

37. Shai Oster, "Illegal Power Plants, Coal Mines in China Pose Challenge for Beijing," *Wall Street Journal*, December 27, 2006.

38. *Kaldaya.net*, "China Unable to Guarantee the Safety of its Nuclear Plants," May 30, 2011.

39. *BBC News*, "China Nuclear Accident Revealed," July 5, 1999.

40. Ibid.

41. Kent Ewing, "Plane Crash Revives China Safety Fears," *Asia Times Online*, September 15, 2010.

42. World Nuclear Association, "Nuclear Power in China."

43. *Want China Times*, "Wenzhou Oversights Reveal Risks of China's Nuclear Program," August 3, 2011.

44. Yuriy Humber, Sangim Han and Shinhye Kang, "Nuclear Industry Says Back on Track After Fukushima 'Speed Bump,' " *Bloomberg Businessweek*, March 25, 2012, http://www.businessweek.com/news/2012-03-25/nuclear-industry-says-back-on-track-after-fukushima-speed-bump.

45. Leslie Hook, "China's Nuclear Freeze to Last to 2012," *Financial Times*, April 12, 2011.

46. David Stanway, "In China the Big Nuclear Question Is 'How Soon?' " Reuters, May 8, 2011.

47. See Nuclear Age Peace Foundation, "China," *Nuclearfiles.org.*; and Nuclear Control Institute, "China's *Non*-Proliferation Words vs. China's Nuclear Proliferation Deeds."

Session III

Economic and Regulatory Issues

The Capture Theory of Regulation 11

GARY S. BECKER

The "capture theory" of regulation usually—although not always—means that regulators act to further the interests of the companies or other groups they are regulating rather than a general public interest. It was pioneered in economics by George Stigler.[1] The theory receives support from studies of the actual policies of regulators in many industries, including the housing industry in the years leading up to the 2008 financial crisis.

The theory does not assume that regulators are corrupt, although some may be, but rather analyzes the interaction between the incentives and behavior of regulators and those of the companies they regulate. Companies can often orient the formulation and implementation of regulations in their favor instead of in the interests of consumers. The reason is that typically the average consumer would only be slightly affected by particular regulations because the effects of the regulations would be spread over millions of consumers. By contrast, the gain to producers from tilting regulations in their favor would be concentrated on a relatively small number of companies, so that the gain per company from favorable regulations would be substantial.

This discrepancy between the behavior of consumers and producers was well stated more than a century ago by Simon Newcomb, an outstanding American economist and astronomer of the nineteenth century. As he put it, "One cent per year out of each inhabitant would make an annual income of $500,000. By expending a fraction of [their] profit, the proposers of policy A could make the country respond with appeals in their favor . . . Thus year after year every man in public life would hear what would seem to be the unanimous voice of public opinion on the side opposed to the public interests."[2]

The general point is that since the number of producers is much smaller than the number of consumers, each producer has a much greater incentive to try to manipulate the formulation and implementation of regulations than does each consumer. In addition, because the number of companies in an industry is relatively small, companies can frequently organize for collective action to induce regulators to act in their interests rather than in the interests of consumers or other groups.

Many examples show how industries twist the implementation of regulations in their favor. In his 1971 article, Stigler described the regulation of interstate trucking prior to its deregulation in the late 1970s. The cumulative number of applications by truckers for certification to transport goods interstate grew by about 80 percent from the mid-1940s to the early 1970s. Yet the total number of licensed carriers declined over this period by about one third. The regulators, members of the Interstate Commerce Commission (ICC), made it virtually impossible for applicants to get certified, while over time some existing carriers went out of business. The ICC used its certification procedures to promote the interests of existing interstate truckers rather than the interests of consumers and companies that were moving goods out of town. Other rules by state as well as federal regulators limited truck size and weight in order to favor the powerful railroad lobby that was fighting the encroachment of trucking on its traditional interstate traffic business.

The deregulation movement of the late 1970s and early 1980s reduced the capture of regulators in airlines, interstate trucking, stock exchanges, and some other industries through reducing the intrusiveness of regulations in these industries. However, regulations increased in labor markets, the housing market, and in other sectors. The regulation of Fannie Mae and Freddie Mac gives a depressingly good example of the capture of regulators by these semi-private companies.

In their recent book, *Reckless Endangerment*, *New York Times* business writer Gretchen Morgenson and financial analyst Joshua Rosner give a detailed discussion[3] of the federal regulation of Fannie and Freddie in the years leading up to and during the financial crisis. The authors show how Fannie Mae used deep political connections and intellectual and personal intimidation to obtain dominant positions in the residential mortgage market. Democratic Massachusetts Rep. Barney Frank and other influential congressmen and government officials defended these companies by claiming that what they were doing was not risky. Yet, at the beginning of the crisis in 2008, Fannie and Freddie held or guaranteed about half of the $12 trillion of assets in the residential housing market. As a result, both companies became insolvent in September 2008, when they were taken over and bailed out by the federal government.

Sometimes, however, active consumer or other groups manage to overcome the powerful forces favoring the industries being regulated. Voting by millions of consumers could have an enormous influence over the directions that regulations take if events induce them to become active. As a result, consumer advocacy groups rather than the companies being regulated may "capture" the regulators in the sense that they induce regulators to act in ways that harm these companies—ways that may even lower, rather than raise, general welfare.

Greater activity by consumer and other groups rather than the companies being regulated is often induced by dramatic events that appear to inflict serious harm on the economy and general welfare due to what is taken to be lax or improper regulation. A good example is the increased requirements for clinical trials and other safety standards that public outrage forced the FDA to introduce after millions of tablets of the drug thalidomide had been distributed to physicians during a clinical testing program. Although the FDA never approved the drug for general use, the tablets that had been distributed to mothers caused some of their children to be born with serious defects. Nevertheless, some economists have argued persuasively (e.g., Sam Peltzman[4]) that the various additional requirements in introducing new drugs resulting from this episode turned out to cause greater harm to sick individuals by the delays and outright prevention of their access to effective medications than the resulting gains from delaying or preventing the approval of drugs that would harm some patients.

Industries where there is a small probability of some very bad consequences from their activities pose a particularly difficult area for regulators. Many drugs, even after they go through rigorous clinical trials, may after widespread use cause a small fraction of the users to suffer serious side effects. The nuclear power industry provides an even better example. Release of radioactive materials in significant quantities from nuclear power plants has been rare, but the damages caused when it happens can be severe. The nuclear power industry may greatly influence the direction of regulations during the prolonged periods when the industry works well. But severe regulatory reactions are triggered by an accident that releases large quantities of radioactive materials, as happened in 1986 at Chernobyl in Ukraine or in 2011 at the Fukushima nuclear plant in Japan due to an enormous earthquake and resulting tsunami. A different example is the accident at the Three Mile Island Nuclear Generating Station in Pennsylvania in 1979. Construction of nuclear power plants in the United States was essentially shut down after that accident. Yet subsequent studies indicate that the Three Mile Island accident caused little health damage to nearby residents.

Stigler, in *The Citizen and the State*,[5] discusses various other dimensions of poor performance by regulators, in addition to their capture by the industries being regulated. He does not believe much can be done to overcome the incentives of regulators to err on the side of delay in approving new drugs or other new and risky products and services. However, more studies like the one by Peltzman cited earlier would help by making more apparent the sometimes major cost of delays in approval. Nor does he see much hope in countering the incentives of companies being regulated to spend lots of time and money trying to influence the content of the regulations and how they are enforced.

Peltzman proposes a few ways to pressure regulators to be more effective in promoting general welfare rather than that of the companies they are regulating. Some government officials should systematically try to evaluate the benefits, costs, and other measures of the success of regulators and regulations. In this way they might change the tendency of regulators to exaggerate their successes, such as by simply counting the number of antitrust cases they won, the amount of illegal drugs seized, or the number of new medical drugs approved. Since Stigler wrote, more proper evaluations have become more common by the Congressional Budget Office, Office of Management and Budget, Government Accountability Office, Council of Economic Advisors, and the White House's regulatory czar (Office of Information and Regulatory Affairs).

Stigler also advocated much greater reliance on private enforcement of various regulations to complement public enforcement. Examples of private enforcement include antitrust civil lawsuits, private product liability lawsuits against companies with allegedly defective products, and malpractice suits against doctors. These types of private enforcement have greatly expanded since Stigler wrote. It is not yet clear by how much they improved the enforcement of regulations since some of these private actions are either frivolous or fishing expeditions by lawyers. Yet, overall, I believe private actions should play a major part in the enforcement of most regulations.[6]

Notes

1. George J. Stigler, "The Theory of Economic Regulation," *The Bell Journal of Economics and Management Science*, vol. 2, no. 1 (Spring, 1971).
2. Simon Newcomb, *Principles of Political Economy*, 1885 (republished by Cornell University, 2009), p. 459.
3. Gretchen Morgenson and Joshua Rosner, *Reckless Endangerment: How Outsized Ambition, Greed, and Corruption Led to Economic Armageddon* (New York: Times Books, 2011).
4. Sam Peltzman, "An Evaluation of Consumer Protection Legislation: The 1962 Drug Amendments," *Journal of Political Economy*, vol. 81, no. 5 (September-October 1973).
5. George J. Stigler, *The Citizen and the State: Essays on Regulation* (Chicago: University of Chicago Press, 1975).
6. Full disclosure: in this recommendation, Stigler relies on our joint paper, "Law Enforcement, Malfeasance, and Compensation of Enforcers," *The Journal of Legal Studies*, vol. 3, no. 1 (January, 1974).

The Federal Regulatory Process as a Constraint on Regulatory Capture

12

JOHN F. COGAN

Introduction

Regulatory capture is the process by which a regulated industry exerts influence or control over a government regulatory agency that is supposed to regulate it. To the public at large, regulatory capture is the exception, not the rule. The public, more often than not, is surprised when the existence of heavy influence by a specific industry on a regulatory agency comes to light. Most economists, on the other hand, regard regulatory capture as the rule. Although the extent of industry influence may vary from one regulatory body to another, it is a rare instance when such influence is absent.

This paper examines the federal regulatory process as a mechanism for reducing the extent of regulatory capture. It then proposes a process improvement that could help further insulate federal independent regulatory agencies in general, and the Nuclear Regulatory Commission in particular, from the prospect of regulatory capture.

Regulatory Capture

Regulatory capture is a process by which a regulated entity, such as a business firm or an organized special interest group, influences a government agency. This influence causes the regulatory agency to act in the industry's interest at the expense of other groups and to the detriment of the public interest. Regulatory capture manifests itself in a host of ways. For example, a regulator may set or approve prices that are above their proper market level; a regulatory agency may reduce competition by restricting entry; or a regulatory agency may relax regulatory burdens to lower industry costs.

The practice of capturing regulatory agencies is as old as the agencies themselves. One of the earliest examples of regulatory capture involves one of the federal government's first major regulatory agencies: the Interstate Commerce Commission. The ICC was established in 1887 to regulate the railroads. Soon after its creation, President Grover Cleveland's attorney general Richard Olney, a former railroad man with Chicago, Burlington, & Quincy Inc., received a letter from the firm's president, Charles E. Perkins, asking Olney if he had plans to terminate the agency.

Olney's letter in response to the executive contains a now famous passage:

> The commission . . . is, or can be made of great use to the railroads. It satisfies the popular clamor for a government supervision of the railroads, at the same time that supervision is almost nominal. Further, the older such a commission gets to be, the more inclined it will be found to take the business and railroad view of things. It thus becomes sort of a barrier between the railroad corporations and the people and a sort of protection against hasty and crude legislation hostile to railroad interests . . . The part of wisdom is not to destroy the commission but to utilize it.[1]

Regulatory capture is rarely as overt today as it was during the days of the railroad monopolies. More often, it is a subtle phenomenon which is often hard to spot at the time it is taking place. Undue influence usually becomes apparent only after the fact, when the cumulative impact of many individual regulatory decisions, taken over a long period of time, is observed. Far too often, the influence comes to light only as a result of a crisis or accident.

This observation gives rise to what I think of as a modern-day regulatory cycle. The cycle is especially applicable to public safety and financial soundness regulations. The first part of the cycle occurs when things are going well in the industry. During this period, as a result of undue industry influence on regulators, standards are steadily relaxed, gradually less and less monitoring occurs, and less and less regulatory compliance takes place. The second part of the cycle occurs when the cumulative impact of the industry's influence leads to a crisis or an accident. Following the crisis, the regulator responds with a spasm of regulations which are often excessive and harmful.

The financial crisis of 2008 and the federal government's reaction to it are good examples of this regulatory cycle in action. As John B. Taylor and Frank A. Wolak noted in their conference paper,[2] increasingly lax monitoring by various federal financial services industry regulators and by the federal overseer of Fannie May and Freddie Mac before 2008 led to excessive risk-taking by financial institutions and, ultimately, to their financial collapse. The spasm of regulations that has followed in the wake of the Dodd-Frank legislation is characteristic of the regulatory cycle's second stage.

As Gary Becker writes in his conference paper,[3] regulatory capture exists because the benefits of regulation that are advantageous to industry tend to be concentrated among a relatively small number of firms in an industry. On the other hand, the costs of such regulation tend to be diffused across a large number of consumers. The gain per company in a concentrated industry from influencing a regulation outcome is large, while the cost of that regulation to any particular consumer is small.

In recent decades, as Becker also writes, declining communications costs have reduced the cost of organizing individuals into groups. The result has been a rise in the power of consumer advocacy groups. Such groups possess the ability to influence regulatory actions and they now serve as an important counterweight to a regulated industry's influence. Thus, in the modern world of regulatory capture, factions or groups representing different sides of the market vie with one another for favorable regulatory outcomes.

The Federal Regulatory Process

There is no sure-fire way to entirely eliminate regulatory capture. But two keys to limiting its extent include (1) a publicly transparent process in which all affected parties have input into a final regulatory decision before its promulgation as law and (2) a requirement that final rules be based on evidence in a way that is designed to minimize the degree of arbitrariness in regulatory decisions.

The United States government, over the course of more than five decades, has established an extensive formal rule-making process to govern all major regulations issued by executive branch agencies. This process, codified in the Administrative Procedure Act (5 U.S.C. 552–553), contains several distinct requirements designed to ensure transparency and limit arbitrariness in the issuance of policy rules. On all major regulations, regulatory agencies are required to provide all affected parties with notification of a proposed rule and an opportunity to submit written comments for consideration by the agency prior to its issuance of a final rule. Specifically, regulatory agencies are mandated to meet this requirement by publishing a "Notice of Proposed Rulemaking" in the Federal Register and by allowing sixty days for the public comment period. The agency must also make all written comments publicly available. Transparency is further enhanced by requiring the regulatory agency to submit the proposed final regulation to both the House of Representatives and the Senate before its effective date. As a final step to ensure transparency, the agency is then required to publish the final regulation in the Federal Code of Regulations.

To limit the degree of arbitrariness of regulatory decisions, the federal regulatory process imposes two requirements on agencies. First, during the public comment period, the agency must refrain from engaging in ex parte communications with affected parties individually without all parties present. In practice, this means that the agency will usually not communicate with any parties during the comment period except through notices made publicly available to all parties. Second, final regulatory actions must be based on either a risk assessment or a cost-benefit analysis.[4]

As many observers of the policymaking process have observed, regulatory agencies which are housed within cabinet departments tend to be more susceptible to regulatory capture than independent agencies. It is argued that a cabinet department's larger mission to advocate or promote an industry's interests is likely to improperly influence regulatory decisions of an agency within its structure. As some conference participants noted, the institutional placement of Japan's nuclear industry's regulator, the Nuclear and Industrial Safety Agency (NISA), may have played a contributing role in Japan's recent nuclear disaster. NISA is housed within the Ministry of Economy, Trade, and Industry (METI), a department responsible for promoting the nuclear power industry.

In the United States, many federal regulatory agencies operate as independent agencies rather than as part of cabinet departments. This independent agency status is an important institutional design feature that can provide regulatory agencies with a degree of insulation from pressures to promote a regulated industry's interest.[5] However, in the federal regulatory process, this institutional design has been accompanied by an unfortunate feature that exposes independent agencies to more regulatory capture pressure.

The federal regulatory process requires regulatory agencies to submit both proposed rules and final rules to the Office of Management and Budget (OMB) for review prior to publication. This mandate ensures that the need for, and effectiveness of, any major rule receives independent analysis and assessment. The OMB review provides a second look at the regulation and serves as a check against regulations that are not adequately justified on the basis of costs and benefits or risks to society at large.

The unfortunate feature of the U.S. regulatory system is that independent regulatory agencies are exempt from this requirement. Independent agencies, neither by law nor practice, submit their proposed or final regulatory actions to OMB for review. For independent agencies, the second look is absent and, as a result, an important constraint against undue industry influence is missing.

This observation leads to a simple, but important, policy recommendation: require all independent regulatory agencies to submit their proposed major regulations to OMB for review. This recommendation may be particularly important for the Nuclear Regulatory Commission. Virtually all of the agency's funding comes from industry contributions. This is likely to make the agency more susceptible to regulatory capture. "He who pays the piper, calls the tune." Although there is little evidence of a significant degree of industry influence, as I noted earlier, evidence often comes too late and in the form of a crisis or an accident. Under such circumstances, instituting an extra form of protection against regulatory capture of the NRC may be especially warranted.

Notes

1. Gerald G. Eggert, *Richard Olney: Evolution of a Statesman* (Philadelphia: The Pennsylvania University Press, 1974), p. 28.
2. John B. Taylor and Frank A. Wolak, "A Comparison of Government Regulation of Risk in the Financial Services and Nuclear Power Industries," chapter 13, this volume.
3. Gary S. Becker, "The Capture Theory of Regulation," chapter 11, this volume.
4. Several conference participants noted that in cases in which outcomes rarely occur, such as nuclear power accidents, there is considerable uncertainty about the probabilities that should be assigned in performing risk assessments or cost-benefit analyses. In these cases, the usefulness of incorporating either risk assessments or cost-benefit analyses as a means of limiting arbitrary regulatory outcomes is itself limited.
5. As the 2008 financial crisis has demonstrated, it is not clear just how much this insulation is worth. The Securities and Exchange Commission, the Federal Reserve, and the Federal Deposit Insurance Corporation are all independent agencies and, as the record of their performance over the last decade has indicated, they are quite susceptible to regulatory capture.

A Comparison of Government Regulation of Risk in the Financial Services and Nuclear Power Industries

13

JOHN B. TAYLOR AND FRANK A. WOLAK

In this paper we examine and compare the problems of safety and soundness regulation in two industries that have dominated the headlines in recent years—nuclear power and financial services. In both industries, a major purpose of regulation and supervision is the prevention of excessive risk-taking which can result in extraordinary harm to society. While clearly important industries in their own right, we hope a side-by-side comparison sheds light on the broader policy problem. One of the most

This is a revised version of a draft prepared for discussion at the Nuclear Enterprise Conference, October 3–4, 2011, Hoover Institution, Stanford University.

striking differences in our comparison is that the recent safety record in the nuclear power industry in the United States appears to be better than the safety record in the financial sector, as evidenced most recently by the failure and bailout of several large financial firms and the resulting global financial crisis.

An important issue in addressing the adequacy of regulation and supervision is the problem of regulatory capture, or the tendency for regulated firms and their government regulators to develop mutually beneficial relationships that are harmful to the economy, public safety, and people's lives more generally. The benefits to the regulated firms may include lax supervision, protection from competition, and even government bailouts. The benefits to the regulators may be lucrative post-government employment, political contributions, and favors to family and friends, which may be implicit or explicit. Of course, the underlying problem of regulatory capture goes beyond heavily regulated industries. Sometimes called the problem of crony capitalism, it can exist in any industry in which government policy has a large role, which is the vast and growing majority of industries today.

This across-industry comparison of the regulation of risk yields insights about which mechanisms from one industry can be usefully transferred to the other industry and which cannot or should not. Moreover, understanding why certain mechanisms cannot be readily transferred suggests directions for future research into the design of mechanisms for regulating risk for both sectors. Although there are a number of reasons for the relative success of the nuclear power sector in managing risk, we focus on those which may have lessons for regulation of risk in the financial sector, including the success of a non-profit industry sponsored organization—the Institute of Nuclear Power Operations (INPO)—which provides comprehensive monitoring and strong incentives to reduce risk.

Changes in the structure of the electric utility industry over the past twenty years have created challenges for the regulation of risk that are similar to those that exist in the financial sector. For this reason, modifications of the current regime of safety regulation in the nuclear power sector may be necessary to address these new sources of conflict between maintaining public safety and the company's financial condition.

Regulating and Supervising Financial Services

The financial sector in the United States is regulated and supervised by a large number of federal and state government agencies with overlapping jurisdictions. Commercial banks are regulated by the Federal Reserve, the Office of the Comptroller of the Currency, and the Federal Deposit Insurance Corporation (FDIC), along with state regulatory agencies. Securities firms and markets are regulated by the Securities and Exchange Commission and the Commodity Futures Trading Commission. Insurance companies are regulated at the state level. The government-sponsored enterprises Fannie Mae and Freddie Mac are regulated by the Federal Housing Finance Agency (FHFA), which combined the previous Office of Federal Housing Enterprise Oversight (OFHEO) with the Federal Housing Finance Board.

The Dodd-Frank Wall Street Reform and Consumer Protection Act of 2010 expanded the Federal Reserve's responsibility beyond bank holding companies to all systemically important financial institutions. It also created a new Bureau of Consumer Financial Protection financed by the Federal Reserve and a Financial Stability Oversight Council to coordinate regulatory agencies across the federal government with the help of a new Office of Financial Research. The Dodd-Frank financial reform bill also created an "orderly liquidation authority" in which large ("too big to fail") financial firms could be resolved by the FDIC rather than go through bankruptcy proceedings. There is disagreement about whether Dodd-Frank increases or decreases the chance of government bailouts of such firms. An alternative, or supplement, to the orderly liquidation authority would be to create a special bankruptcy process for financial firms called Chapter 14.[1]

Rule-making is an important part of financial market supervision and regulation. For example, the Dodd-Frank bill alone requires 235 different rule-makings in which broad requirements in the legislation are translated into specific regulations. Most of the regulatory agencies listed above are involved in the rule-making process.

Rule-making proceeds in accordance with the Administrative Procedure Act (APA) of 1946 in which agencies draft rules which are then put out for public comment and then revised. The final rule is then published in the Federal Register. The financial industry, including trade groups like the Securities Industry and Financial Markets Association (SIFMA), is of course deeply involved in rule-making. While it is essential to draw on industry expertise, this raises another route for lobbying for special treatment.

Global financial firms are also regulated by government agencies in other countries. Some international coordination is provided by groups like the Financial Stability Board, which has recently designated twenty-nine large banking firms as "systemically important." They include twelve headquartered in the United States, thirteen in Europe, and four in Asia. The designated firms are Bank of America, Bank of New York Mellon, Citigroup, Goldman Sachs, J.P. Morgan, Morgan Stanley, State Street, Wells Fargo, Royal Bank of Scotland, Lloyds Banking Group, Barclays, HSBC, Credit Agricole, BNP Paribas, Banque Populaire, Societe Generale, Deutsche Bank, Commerzbank, Unicredit, UBS, Credit Suisse, Dexia, ING, Banco Santander, Nordea, Mitsubishi UFJ, Mizuho, Sumitomo Mitsui, and Bank of China.

Excessive Risk-Taking and the Financial Crisis

The financial crisis and panic of 2008 revealed serious deficiencies in the regulatory and supervision process. Although there is disagreement about whether the scope of regulation was too narrow, there is little disagreement that the rules and regulations already in place were not adequately enforced by the regulators and supervisors in a number of important cases.

The most documented cases are Fannie Mae and Freddie Mac, the two giant private, government-sponsored enterprises that support the

U.S. housing market by buying mortgages and packaging them into marketable securities which they then guarantee and sell to investors or add to their own portfolios. Both organizations have been regulated by government agencies with the purpose of preventing excessive risk-taking by the institutions.

However, by any reasonable measure, both agencies undertook excessive risk starting in the late 1990s by guaranteeing home mortgages which had a high probability of default. For this reason they share significant blame for the financial crisis and the global recession. By encouraging people to take on such loans they accentuated the housing boom which led to the financial bust. Obviously their regulators failed in their most fundamental responsibility. The result was catastrophic.

But Fannie Mae and Freddie Mae were not the only regulated financial institutions that took on excessive risk. Large financial firms from Citibank to AIG to Bank of America to Bear Stearns to Lehman Brothers—all regulated by government agencies with the stated purpose to prevent excessive risk—were heavily invested in risky securities based on questionable home mortgages or other debt. The regulatory system apparently failed as well.

Regulatory Capture: The Case of Fannie Mae

What role did regulatory capture have in causing this failure? In the case of Fannie Mae and Freddie Mac, the evidence of regulatory capture is substantial. Gretchen Morgenson and Joshua Rosner[2] document a massive support system between well-connected individuals in these organizations and government officials which encouraged excessive risk-taking. By providing favorable regulatory treatment and protection from competition, the government enabled Fannie Mae to bring in $2 billion in excess profits, according to a 1996 study by the Congressional Budget Office.[3]

Morgenson and Rosner provide considerable information about how the regulatory capture took place. Fannie Mae officials got jobs for friends and relatives of elected officials. They set up partnership offices around the country which provided more such jobs. They financed publications in which writers argued that Fannie's role in promoting home-ownership justified federal support. They commissioned work by famous economists who argued that Fannie was not a serious risk to taxpayers. In this way, they countered critics who argued that both Fannie and Freddie posed significant risks.

The officials made campaign contributions and charitable donations to co-opt groups like ACORN, which had been asking for tighter regulations. They persuaded executive branch officials to ask their staffs to rewrite reports critical of Fannie. The mortgage firm Countrywide partnered with Fannie in originating many of the mortgages Fannie packaged (26 percent in 2004) and gave "sweetheart" loans to politicians with power to affect Fannie.

Fannie's lobbying efforts were resisted by some government officials. Then CBO Director June O'Neill, for example, refused to stop the release of the 1995 CBO study showing that federal support increased Fannie's profits by $2 billion. Then Secretary of Treasury John W. Snow proposed in 2003 the creation of a new federal agency to regulate and supervise Fannie.

While this problem was most severe in the case of Fannie Mae and Freddie Mac, the same problem also apparently existed between some of the large Wall Street financial firms and their regulators at the Federal Reserve Bank of New York. It is very hard to imagine that heavily regulated banks could have engaged in such extreme risk-taking without at least the implicit support of regulators, and when Citigroup built up its hidden off-balance-sheet risks in 2006, the New York Fed did not object, though it had the power to do so. However, the same type of hard evidence that Morgenson and Rosner found in the case of Fannie Mae and Freddie Mac has yet to surface in the case of the other firms.

Firms in other industries undoubtedly take actions to increase their influence and receive favorable regulatory treatment. But what is different in the case of Fannie Mae, Freddie Mac, and the other financial firms is that the government responded with such lax regulatory oversight and favorable treatment that the entire economy suffered severely.

Regulatory failure in the nuclear sector has economy-wide impacts similar to the financial sector, particularly for the regions near a nuclear disaster. However, as we discuss below, a single safety disaster in one firm threatens the financial viability of all firms in the nuclear sector. This provides incentives for firms to take actions to discourage excessive risk-taking at other firms in the nuclear sector, an incentive which has apparently been absent in the financial sector. In addition, the potential profitability upside to excessive risk-taking in the nuclear power sector is limited because the revenues received by plant owners are regulated by government agencies. However, the regulation of revenues that nuclear generation unit owners receive is light-handed for generation units selling their output in bid-based wholesale markets, because competition is thought to be sufficient to discipline the prices these generation unit owners receive.

Regulating Nuclear Power

Nuclear power in the United States is subject to federal regulatory oversight for safety and output price by two government agencies, the Nuclear Regulatory Commission (NRC) and the Federal Energy Regulatory Commission (FERC). The safety oversight function of the NRC—which has an overall annual budget of $1 billion and employs about 4,000 people—includes nuclear power plant commissioning licensing for the use of radioactive material and waste repositories and monitoring of operations. Both the NRC and FERC employ administrative procedures established under the APA for regulatory decision-making. The power of these independent regulatory agencies (IRAs) was established through enabling legislation that allows them significant scope for the interpretation of legislative intent in carrying out their regulatory mission.

The framework set out by the APA establishes a quasi-legal process managed by an administrative law judge (ALJ) to conduct public hearings and make decisions with respect to power plant commissioning for the case of the NRC and output pricing for the case of the FERC. Interested parties with standing on an issue are allowed to submit evidence to the public proceeding and can cross-examine witnesses offered by other parties. The ALJ's decision following the conclusion of the evidentiary hearings, along with all of the accompanying evidence and testimony presented, is forwarded to the members of the regulatory commission, which is free to adopt the ALJ's decision or write its own. However, any decision the regulatory commission reaches must follow from the evidence presented in the ALJ process only, so that it is most common for the ALJ's decision to be adopted with only minor modifications.

In making its final decision on an issue, the commission is not allowed to rely on other information, such as private communication with one of the parties. In fact, private contact between commission members and parties on a pending issue before the commission, what is typically called ex parte communication, is prohibited. Violation of this rule is grounds for legal review of the commission's decision. The other basis for legal review of a commission decision is a failure to make the decision in a manner that is consistent with the intent of the enabling legislation. Because of the technical nature of many of the decisions that come before these two commissions, the courts give significant deference to them in interpreting the intent of the law.

Potential for Regulatory Capture in Nuclear Power Sector

In many ways, the APA process is designed to limit the scope for regulatory capture. All information relevant to the decision made on a nuclear power plant licensing decision (by the NRC) or a pricing decision (by the FERC) must be presented in a public hearing and be subject to cross-examination. Any decision ultimately made by the regulatory commission is also subject to judicial review for consistency with the intent of the underlying legislation and consistency with due process, i.e., the decision follows directly from the evidence presented.

However, there are a number of details of the regulatory processes at these two commissions that may allow scope for regulatory capture. First, the ALJ process can be extremely costly and time-consuming relative to a more informal process. The somewhat open-ended nature of the evidentiary process can allow a participant with deep pockets, such as a nuclear power plant owner, to impose significant costs on other participants in the regulatory proceedings. Therefore, well-financed participants may be able to exert a greater influence on the outcome of the regulatory process.

Some regulatory agencies have attempted to address this issue by allowing certain classes of interested parties to recover their costs of participation. For example, environmental groups participating in a nuclear power plant licensing process may be able to file with the commission to recover the costs of their participation from the owner of the actual or proposed power plant.

A second issue concerns the selection and monitoring of administrative law judges. Few of the regulatory agencies collect and disclose systematic information on the decisions of their administrative law judges. These judges are also employees of the agencies and may therefore have career concerns both in and outside the agency that can affect their decisions. There is a rigorous and relatively independent process for selecting lawyers to become ALJs that is designed to reduce the opportunities for regulatory capture. However, for a variety of reasons, most agencies typically do not use this process for selecting ALJs. A procedure called "selective certification" allows a regulatory agency to bypass this formal process and choose its own candidate from the set of candidates for an ALJ position that the independent process has determined to be "qualified." The upside of the selective certification approach is that the regulatory agency can appoint someone with the necessary technical expertise for that specific regulatory agency to serve as an ALJ. The downside is that the selective certification process opens up the opportunities for regulatory capture.

A crucial issue that limits the scope for regulatory capture is the recognition following the Three Mile Island (TMI) accident in 1979 that the current and future financial viability of the entire nuclear power industry depends on there being no nuclear accidents. As Detroit Edison CEO Walter J. McCarthy Jr. stated in a speech to other nuclear utility CEOs at the time, "Each licensee is a hostage of every other licensee" and "we truly are all in this together."[4] This perspective led the industry to form INPO as an industry-run regulatory bureaucracy to ensure the safe and reliable operation of nuclear facilities in the United States. As we note below, INPO has been a major factor in achieving the significant safety and reliability improvements of the nuclear power sector since TMI.

Because there are ready alternatives for producing electricity besides nuclear power, the entire nuclear industry faces the risk that all nuclear power plants will be shut down and no further power plants built if there is an accident. The recent discussions in Japan and many other countries about the future of nuclear power and the decision of several European countries to abandon nuclear power in response to the Fukushima disaster justify McCarthy's statements. In this sense, the nuclear power sector may have a built-in incentive for strong self-regulation of catastrophic risk which does not currently exist in the financial sector. While there has been a political backlash to the recent actions of Wall Street financial firms and their regulators, the view that the financial sector is essential to the economy limits adverse effects on the entire industry.

Performance of U.S. Nuclear Power Sector

The performance of the United States' nuclear power industry has been increasingly impressive on both safety and economic grounds since the 1970s. The annual capacity factor of a generation unit is a useful summary measure of the operating efficiency of a nuclear power plant because the low variable cost of operating nuclear facilities implies that, except in extremely rare circumstances, if a generation unit is available to operate, it will operate. The annual capacity factor of a generation unit is defined as the total amount of output produced by the generation unit annually divided by the nameplate capacity of the generation unit times the total number of hours in the year. For example, if a 50 megawatt (MW) generation unit produced 328,500 MW-hours (MWh) annually, it would have an annual capacity factor of $0.75 = 328{,}500$ MWh/(50 MW $\times$ 8760 hours).

Note that for other power plants, a capacity factor may not be the most useful measure of operating efficiency. For example, a natural gas-fired power plant may be too expensive to operate during certain hours of the year simply because demand is not sufficiently high to require the unit to operate. Consequently, a lower capacity factor for a natural gas-fired unit does not imply it is unable to operate in fewer hours of the year than a nuclear generation unit.

The fleet-level average capacity factor is a summary measure of the operating efficiency of the United States nuclear power plant fleet. A higher value of this capacity factor implies that more electricity is being produced from the same amount of installed nuclear generation capacity. To compute the fleet-level average annual capacity factor, simply repeat the calculation described above with the numerator equal to the annual output of all nuclear generation units in the United States and denominator equal to the sum of total nuclear generation capacity in the United States times the number of hours in the year.

Figure 13.1 plots the fleet-level average annual capacity factor for nuclear generation units in the United States from the early 1970s until now. The figure starts at less than 50 percent in the early 1970s and steadily rises to more than 91 percent in 2010. It is important to note that although no new nuclear power plants were completed after 1996, total nuclear capacity has increased by a few percentage points per year since then because of capacity additions at existing facilities. Nevertheless, fleet-level

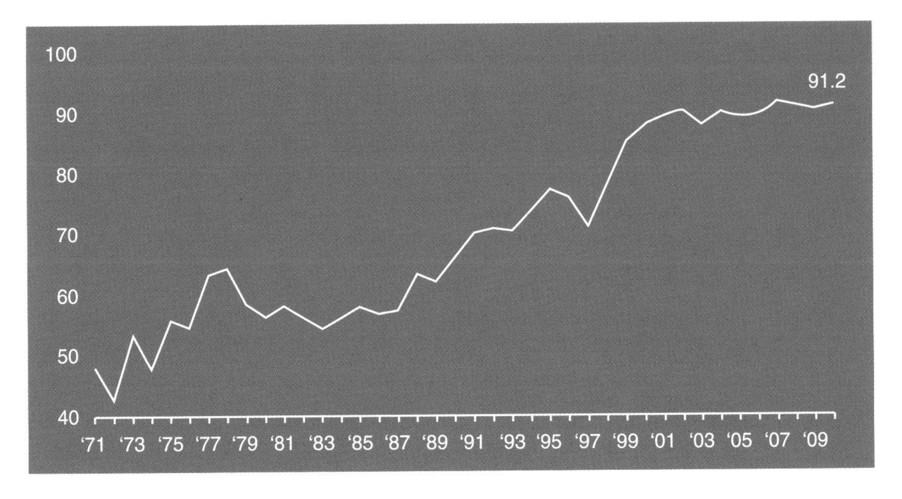

Figure 13.1 United States Fleet-Level Nuclear Capacity Factors (1971–2010, Percent)
Source: Courtesy of Nuclear Energy Institute, based on data from the U.S. Energy Information Administration.

capacity factors continued to rise steadily after that date. A number of explanations have been offered for this tremendous improvement in operating efficiency in the past thirty years.

On the nuclear safety side, there have been no major nuclear power plant incidents in the United States since Three Mile Island in 1979. The hazard of an unplanned outage, the instantaneous conditional probability of an outage given that the plant has not had an unplanned outage up to that point in time, has declined since TMI.[5] This result implies a lower rate of unplanned outage events per plant-year during the post-TMI time period. The reduction in the risk of an unplanned outage in the post-TMI period is largest for plants with the highest risk of an unplanned outage during the pre-TMI time period.[6] In addition, although there has been a continuous reduction in the hazard of an unplanned outage at all nuclear power plants since the time each plant began operating, the rate at which the unplanned outage hazard rate declined has accelerated during the post-TMI period.[7]

A broad measure of overall plant safety and performance compiled by INPO for the entire U.S. nuclear power plant fleet has shown steady improvement over the post-TMI time period.[8] This index is based on measures of power plant performance such as the generation unit capacity factor and safety measures such as the forced loss rate (the percentage of energy generation during non-outage periods that a plant is not capable of supplying because of unplanned energy losses) and scrams (automatic shutdowns) per 7,000 hours of operation. Over the ten-year period from 1995 to 2004, these average performance index measures almost doubled.[9]

Explaining Nuclear Reliability and Safety Improvements

There are a number of factors driving these increases in reliability and safety. The first is the increasing standardization of operating procedures in the nuclear industry driven primarily by the formation of INPO and the changes in regulatory oversight at NRC in the aftermath of TMI. This standardization was facilitated by the Electric Power Research

Institute (EPRI) which had been earlier set up voluntarily by the utilities to sponsor research, including research on ways to improve safety. With all U.S. nuclear utilities participating, EPRI worked with engineers from the utilities to develop standardized safety guidelines and do research on inspection technologies to detect potential failures and identify aging deterioration. The second factor is the consolidation of plant ownership in the nuclear power industry, with a smaller number of firms owning and operating nuclear power plants according to a common set of operating procedures. A third factor is the increasing use of performance-based regulatory schemes that reward the generation unit owner with a higher rate of return for higher capacity factors and punish it with a lower rate-of-return for lower capacity factors. The fourth factor is the introduction of formal wholesale electricity markets where plant owners sell the output they produce at a market-clearing price rather than under a regulatory process that only allows the plant owner the opportunity to recover all prudently incurred costs associated with constructing and operating the nuclear power plant.

Institute of Nuclear Power Operations

INPO was founded in December 1979 as a not-for-profit organization by the nuclear power industry in response to the TMI accident. INPO's mission is: "To promote the highest levels of safety and reliability—to promote excellence—in the operation of commercial nuclear power plants."[10] This is accomplished through nuclear power plant evaluations that assess: (1) the knowledge and performance of plant personnel, (2) condition of systems and equipment, (3) quality of programs and procedures, and (4) effectiveness of plant management.[11] INPO also runs the National Academy of Nuclear Training to provide training for nuclear power professionals and evaluate the quality of individual plant and utility training programs. INPO provides reviews of significant events in the nuclear power industry and shares the lessons learned and best practices throughout the industry. Finally, INPO provides assistance with specific technical or management issues at the request of an individual plant owner.

James Ellis, the CEO of INPO, attributes INPO's success as a self-regulating industry group to five factors: (1) CEO engagement, (2) nuclear power safety focus, (3) support from the nuclear industry, (4) accountability, and (5) independence. There is general agreement that the CEO engagement is the primary driver of the success of INPO.

From the beginning the INPO's board of directors has been composed of CEOs of companies that own nuclear power plants. INPO provides briefings personally to the CEO of each operating company in the presence of the company's management on INPO's performance evaluations (also called on-site peer reviews) conducted at the plant. INPO also holds annual meetings with all of its members where power plants are graded on their performance and these grades are shared with all of the CEOs in an executive session. A CEO of a nuclear power company provides the following description of the process:

> All the CEOs are gathered in a big room with Zack Pate [INPO's president at the time], and he flashes up the most recent evaluation numbers for each of the utilities by name. That's the only time we learn how our peers are ranked, and it kind of hits you right between the eyeballs. The first slide has all the number ones, the best-rated utilities. Lots of praise from Zack, and all those CEOs kind of puff up and get a big smile on their face. [They also receive a plaque.] Then come the number twos, and those guys also feel pretty good about it. And then come the number threes, and they just kind of sit there passive. Then you get down to the fours and the fives. And after some pretty frank discussion of their problems, those guys are feeling rather uneasy to say the least. I guess you could say it's a sense of pride or ego. All CEOs are pretty egotistical. I mean these are people who have worked their way up to managing a major utility, and our societal cultural aim is to strive to be the best and get to the top of the pyramid. I think that's really the driver here. We all want to be a one, and none of us want to be viewed as a poor performer among our peers.[12]

This "management by embarrassment," as this CEO referred to it, has been very effective at causing the owners of the laggard power plants to improve their performance.

Nuclear safety focus means that safety has been the exclusive focus of INPO's activities in spite of calls for INPO to become involved in other issues that relate to the nuclear power industry. *Support from industry* means that the industry understands and accepts that it must subject its plants to on-site peer reviews. Since 1980, INPO has conducted more than 1,200 such reviews, an average of more than sixteen at every nuclear power plant.[13] *Accountability* implies that INPO can be confident that its recommendations are implemented in a timely fashion. Over time, INPO has increased its ability to make nuclear power plant owners accountable in an unconventional manner. Originally, INPO's plant evaluations were distributed industry-wide in an effort to publicly shame poor performance. However, this led to the following unintended outcome described by Joseph Rees:

> After reading dozens of these reports, one can't help noticing their carefully restrained wording and their tactfully diplomatic tone. Nor can one avoid the sense that, for some reason, INPO officials were extremely reluctant to use bluntly candid language in their written assessment of a nuclear plant's performance. They were pulling their punches, and the reason why is not hard to uncover. As an industry organization, INPO was responding to the concerns of its most powerful constituency—the nuclear utility CEOs—who were understandably nervous about the risks associated with such reports.[14]

As a consequence, INPO changed to distributing the evaluation reports only to the plant's owner, including the company CEO. James Ellis, CEO of INPO, said this change "provided more open and candid interactions and discussions of problems or areas for improvement. The confidentiality of reports has proven to be an important aspect of performance improvement and nuclear safety."[15]

INPO also has the ability to impose sanctions on power plant owners, even though it does not have the statutory authority to shut down an operating plant. That power resides with the NRC. However, on several occasions INPO has exerted pressure on power plant owners to shut down plants, delay starting up a plant, or even change the company's management if a safety issue were not properly addressed. Joseph Rees describes the example of Philadelphia Electric's Peach Bottom plant as an example of INPO's ability to enforce its recommendations in the face of significant resistance from the plant owner.[16]

An initial INPO evaluation of the Peach Bottom plant found that the management was unwilling to take the "appropriate actions to stress and enforce standards of expected performance." INPO also perceived a "lack of corporate support in implementing needed changes." Finally, INPO found that "long-standing company practices" were an impediment to implementing the necessary changes. A follow-up evaluation twelve months later again found serious problems at Peach Bottom. Ongoing interactions between INPO staff and the Peach Bottom operators finally resulted in INPO's president writing a stern letter to Philadelphia Electric's CEO and having a private meeting with the CEO and company executives. A follow-up evaluation still found a large number of problems at the plant.

This plant owner management versus INPO stalemate was finally resolved when INPO formed an industry panel to respond to the Peach Bottom problem. This panel was harshly critical of Philadelphia Electric's senior management and ultimately led to its board of directors causing the "early retirement" of the top managers at the company. The new management at Philadelphia Electric subsequently implemented INPO's reforms at the Peach Bottom plant and safety and reliability improved. This sequence of events is often cited by industry observers as the first example of INPO's ability to make the industry accountable for its recommendations.

The final factor, *independence*, has been somewhat of a challenge for INPO. Initially, the founders of INPO had the idea that members of the board of directors should be from outside of the nuclear power industry to ensure its independence. However, the argument that such a board would lack legitimacy from the industry and therefore reduce the effectiveness of INPO as a regulator ultimately led to a board composed of industry executives. INPO is also independent of the NRC, although a number of commentators have argued that the NRC defers much of its industry regulation and oversight functions to INPO and often adopts INPO standards as its own standards.

Industry Consolidation

According to the World Nuclear Association, as of the end of 1991, 101 utilities had some ownership interest in nuclear power plants.[17] At the end of 1999, that number had dropped to eighty-seven, and the largest twelve utilities owned 54 percent of industry capacity. Today, ten utilities own more than 70 percent of industry capacity in the United States. This consolidation has come about through mergers as well as sales of individual facilities. There has also been consolidation in the number of operators of nuclear power plants, with forty-five operators in 1995 dropping to twenty-five today.

Most of the nuclear generation capacity consolidation was the result of mergers. The merger of Unicom and PECO in 2000 formed Exelon, which created the largest nuclear energy supplier in the United States. Exelon has ten nuclear plants that produce roughly 20 percent of U.S. nuclear electricity output. In 2000, Carolina Power & Light merged with Florida Progress Corporation to become Progress Energy, which owns five plants in North Carolina, South Carolina, and Florida. In 2001, First-Energy Corporation merged with GPU Inc.

Management contracts have also been used to consolidate plant operations. Companies that own multiple generation units typically form management companies that contract to provide operating and maintenance services to companies that own single generation units. For example, Exelon currently has management contracts with PSEG of New Jersey to operate nuclear units that it owns. Exelon markets the "Exelon way" of nuclear power plant operation and maintenance as leading to lower variable operating costs and higher plant-level capacity factors than other plants in the industry.

The motivation for both types of consolidations is to capture cost savings and operating efficiencies. A number of studies have found evidence of "learning by doing" in nuclear power plant operations, at both the industry level and utility level.[18] In fact, one study comparing learning by doing in France versus the United States finds that a sizable operating efficiency penalty was paid by the U.S. nuclear power industry as a result of the many different technologies used by different regional utilities and the lack of multi-plant ownership of nuclear generation facilities during the early stages of the nuclear industry.[19] In contrast, France, with its standardized plant designs and multi-plant ownership, experienced significantly higher operating efficiencies much earlier in its deployment of the nuclear generation technology.

Incentive Regulation Schemes

The goal of incentive regulation schemes is to make it profitable for price-regulated nuclear power plant owners to reduce costs by achieving high capacity factors. A popular example of this sort of scheme was the Diablo Canyon Performance-Based Pricing plan adopted in 1988 between Pacific Gas & Electric (the plant owner) and the California Public Utilities Commission (the regulator), which fixed the price at which energy produced by Diablo Canyon was sold. This price was set significantly above the variable operating cost of a Diablo Canyon unit, so PG&E had a strong financial incentive to produce as much output as possible from this power plant. Other state regulators enacted similar incentive regulation schemes for the nuclear power plants under their jurisdictions, and have achieved similar results. These incentive regulation schemes also provided strong incentives for the plant owners to reduce the operating and maintenance costs associated with their facilities. A number of researchers have documented significant operating cost savings for nuclear power plants over this time period.[20]

Electricity Industry Restructuring

Electricity industry restructuring, where nuclear generation unit owners face the default option of selling their output into a formal bid-based wholesale market, provides strong incentives for owners to maximize the capacity factor of their generation units. Typically, the market-clearing price in the wholesale market is set by generation units with significantly higher variable operating costs, particularly during the high demand periods of the day. Consequently, a nuclear power plant owner can earn significant revenues in excess of its variable costs by operating during hours of the day when the market price is above the variable operating costs of the generation unit. A number of studies have documented incremental increases in annual average capacity factors associated with nuclear generation units selling into a restructured market relative to selling into a vertically integrated regulated market structure.[21]

The Potential for Future Regulatory Failure in the Case of Nuclear Power

The recent outstanding safety and performance record of the United States' nuclear power sector is no reason for complacency. There have been no plant retirements since 1998 and most of the plants in the existing fleet have either received or expect to receive twenty-year extensions on their licenses from the NRC. These extensions would bring the projected operating life of most plants to almost sixty years, which is significantly longer than was envisioned at the time these plants were built.

The potential risk of future regulatory failure is the result of the combination of an aging nuclear fleet and the fact that an increasing number of facilities sell into formal wholesale electricity markets. This circumstance can potentially increase risks due to the potential for large profits if the generation unit owner is willing to engage in privately risky behavior (continuing to operate in spite of potential safety concerns at the nuclear facility unit because of high wholesale prices) that can have publicly harmful consequences (creating a catastrophic failure). For example, the average wholesale prices during 2010 to mid-2012 in most U.S. wholesale electricity markets averaged in the range of $50/MWh to $60/MWh. However, during periods of stressed system conditions, prices can hit as high as $3,000/MWh, which was the case during the summer of 2011 in the Electric Reliability Council of Texas (ERCOT) wholesale electricity market.

With variable operating costs for the nuclear power plants in the range of $15/MWh, a nuclear power plant owner can earn massive variable profits by producing as much output as possible during these high-priced periods. This creates an incentive for the unit owner to produce as much output as possible during these periods, which may lead the owner to produce energy from a facility that would be shut down for safety reasons during a period with lower wholesale prices. Consequently, the combination of high-powered incentives to keep operating costs down, aging nuclear facilities, and the potential for sustained periods of high wholesale electricity prices could increase risks of a catastrophic nuclear power plant failure.

It is important to note that the incentive for a catastrophic failure is lower in the case of a vertically integrated monopoly regime where plant owners face cost-of-service price regulation. In that case, generation units would only be able to recover their prudentially incurred costs regardless of operating conditions, including times of stressed system conditions. This lack of a financial upside, and the significant financial downside associated with a catastrophic plant failure, makes it unlikely that a price-regulated expected profit-maximizing plant owner would push a plant beyond its safe operating limits or economize on operating expenses, as long as these operating expenses were deemed to be prudently incurred by the regulator.

In sum, the possibility of an industry-wide catastrophe is remote because the price of wholesale electricity typically differs by location and hour of the day. Therefore, it would be extremely rare for all nuclear power plants in the United States to experience a sustained period of extremely high wholesale prices.

Conclusion

In this paper we compared risk regulation in the financial sector and in the nuclear power sector in the United States. Both sectors are heavily regulated and both are susceptible in principle to regulatory capture. However, the safety record in the nuclear power sector is much better than the record in financial services in the United States, as evidenced by the recent severe financial crisis and the lack of a major nuclear incident since Three Mile Island. In effect, regulatory failure—including through regulatory capture—has been much more of a problem in the financial industry.

There are many differences between these two industries, including the competitive structure, the availability of substitutes, the importance of proprietary information, the ability to monitor risks, and the degree of price regulation as distinct from risk regulation. Nevertheless, there are a number of useful lessons from the successful regulation of catastrophic risk in the nuclear power sector that could improve the regulation of risk in the financial sector. In particular, study of the feasibility of a financial industry analog to the Institute of Nuclear Power Operations would be a worthwhile topic for future research.

Acknowledgments

We are grateful to Burton Richter and John J. Taylor for their helpful comments on a previous draft and to Charles Calomiris, Allan Meltzer, Ken Scott, and participants at the conference for useful discussions.

Notes

1. See Kenneth Scott, George Shultz, and John B. Taylor, eds., *Ending Government Bailouts As We Know Them* (Stanford: Hoover Press, 2010).

2. Gretchen Morgenson and Joshua Rosner, *Reckless Endangerment: How Outsized Ambition, Greed, and Corruption Led to Economic Armageddon* (New York: Times Books, 2011).

3. Congressional Budget Office, "Assessing the Costs and Benefits of Fannie Mae and Freddie Mac" (Washington, D.C.: U.S. Government Printing Office, 1996).

4. Joseph V. Rees, *Hostages of Each Other: The Transformation of Nuclear Safety since Three Mile Island* (Chicago: University of Chicago Press, 1994), p. 2.

5. Paul A. David, Roland Maude-Griffin, and Geoffrey Rothwell, "Learning by Accident? Reductions in the Risk of Unplanned Outages in U.S. Nuclear Plants After Three Mile Island," *Journal of Risk and Uncertainty*, vol. 13, no. 2 (1996), pp. 175–198. They estimate proportional hazards models for the length of spells of operating hours for all U.S. nuclear power plants and find the hazard rate for an unplanned outage fell significantly after TMI.

6. Ibid, p. 193.

7. Ibid.

8. Andrew C. Kadak and Toshihiro Matsuo, "The Nuclear Industry's Transition to Risk-Informed Regulation and Operation in the United States," *Reliability Engineering & System Safety*, vol. 92, no. 5 (2007), pp. 609–618. They compute the average of the INPO performance index for various types of U.S. nuclear power plants for the period 1995 to 2004.

9. Ibid, p. 613.

10. "Our History" section of INPO Web site, http://www.inpo.info.

11. "What We Do" section of INPO Web site, http://www.inpo.info.

12. Rees, *Hostages of Each Other*, pp. 104–105.

13. James O. Ellis, "Testimony to National Commission on the BP Deepwater Horizon Oil Spill and Offshore Drilling," August 25, 2010.

14. Rees, *Hostages of Each Other*, p. 94.

15. Ellis, "Testimony," August 25, 2010.

16. Rees, *Hostages of Each Other*, p. 110.

17. World Nuclear Association, "Nuclear Power in the USA," updated March 2012, http://www.world-nuclear.org/info/inf41.html.

18. Richard K. Lester and Mark J. McCabe, "The Effect of Industrial Structure on Learning by Doing in Nuclear Power Plant Operation," *The RAND Journal of Economics*, vol. 24, no. 3 (1993), pp. 418–438; and Martin B. Zimmerman, "Learning Effects and the Commercialization of New Energy Technologies: The Case of Nuclear Power," *Bell Journal of Economics*, vol. 13, no. 2 (1982), pp. 297–310.

19. Lester and McCabe, "The Effect of Industrial Structure."

20. For a recent example, see Fan Zhang, "Does Electricity Restructuring Work? Evidence from the U.S. Nuclear Energy Industry," *The Journal of Industrial Economics*, vol. 55, no. 3 (2007).

21. Ibid.

Discussion Notes on the Economics of Nuclear Energy

14

MICHAEL J. BOSKIN

As the world endeavors to meet a growing demand for energy, amid related environmental and national security concerns, the role of nuclear power is increasingly debated. In the United States, there has not been a new nuclear power reactor started since 1977.[1] The United States gets 20 percent of its electricity from nuclear power, far less than some countries. In France, for example, the portion is 79.9 percent. Other countries (e.g., China) have made huge commitments to nuclear power. Meanwhile, expertise in current nuclear power technology has been migrating from the United States to other countries. When viewed in the broader perspective of total global energy, even with robust assumptions on growth in the developing world, the nuclear power share is projected to grow from the current 6 percent to 8 percent. That's non-trivial, important even—but not by itself *the* answer to the many questions concerning energy and the environment.

Why don't we rely completely on private markets, sans government interference, in the supply and demand for energy? And where does nuclear power fit in the answer to that question? In the energy sector, there are cartels (e.g., OPEC), externalities (environment, security), peak load issues, and immense government tax, subsidy, and regulation policy interventions that generate deviations from the perfectly competitive private market model in which private marginal cost pricing occurs automatically and is socially optimal.

Among the many important features of nuclear power stressed by economists is that some costs are not automatically internalized in competitive free-market pricing. These include security, safety, waste disposal, potential accidents, and risk of proliferation and terrorism. For example, private insurance and reinsurance cover only a small part of the risk of the liability associated with accidents. The bulk is implicitly covered by government. The potential scale of losses simply overwhelms the ability of private insurance to credibly cover the losses. Much like the need for massive government response to augment private response in severe disasters (tsunamis, earthquakes, major hurricanes), nuclear accidents, as the Fukushima example demonstrated, are no different if the initial problem is not quickly contained. So governments, i.e., taxpayers, explicitly or implicitly backstop some of the losses and the costs of recovery and rebuilding.

The design of proper incentives in the "too terrible to ignore" disaster scenario is a key policy issue, much as is "too big to fail" in financial regulation. Regulatory capture issues are discussed by Gary Becker[2] and by John Taylor and Frank Wolak.[3] To be sure, the risk (financial and reputational) of bearing large uninsured losses does create incentives for private action. For example, ExxonMobil, Chevron, and ConocoPhillips have created and funded a company to pre-position equipment and people to more rapidly respond to a major oil spill in the Gulf of Mexico. Large risks, from technical to political to commercial, are routinely shared in the oil industry through joint ventures among several companies and in the design of contracts with host governments.

In addition to the promise of readily available electricity, the nuclear power debate focuses on two additional issues. Nuclear power ranks high on the list of potential energy sources, especially if substituting out coal-fired power plants, to reduce greenhouse gas emissions. But in addition to the risk of substantial environmental disaster—witness Chernobyl and Fukushima—it raises issues of how to dispose of nuclear waste (a contentious, thus far unsolved political battle in the United States) and how to deal with an increasing risk of nuclear material getting into the hands of terrorists.

It is important to note that the list of possible "government failures" in addressing potential externalities is not limited to regulatory capture. Taxes, fines, subsidies, and mandates can be too large as well as too small, and the harm is symmetric. Of course, some of the externalities generated by nuclear power and other fuels span national borders: nuclear accidents, carbon emissions, and terrorism, to name a few. That doubly challenges the design, incentives, and accountability of policy responses; witness the corruption in the carbon "offsets" programs. Fortunately, there are important examples of successful responses to environmental challenges. Examples are the global Montreal protocol for elimination of ozone-depleting chlorofluorocarbons and the U.S. sulphur dioxide emissions trading program to reduce acid rain.

A major expansion of nuclear power in the United States must overcome a variety of commercial issues that currently keep it from being cost-competitive. These include high capital costs, financing, insurance, and litigation risk. Just as important, but too little noted, is a point Steve Bechtel impressed on me in a discussion of the nuclear industry: America has basically lost an entire generation of nuclear engineers and scientists. It will be no small or short-term matter to rebuild the human capital to operate a large expansion of nuclear power, let alone leapfrog the design and R&D business. In the interim, some expertise can be imported.

The cost economics of nuclear power plants is driven by many factors and hence depends on a variety of assumptions. These include the costs of alternatives, especially coal and gas; high capital costs, possibly rising after the Fukushima disaster; financing for the high capital costs; decommissioning and nuclear waste storage costs; possible pricing of carbon; insurance; and uncertainty over which risks will be borne by consumers in regulated pricing, by taxpayers, or by producers. Cost comparisons with alternative energy sources are particularly difficult because of such issues.

Figure 14.1 presents projections by ExxonMobil of the cost of alternative sources of electricity generation in 2025 under two scenarios for carbon pricing: zero or $60 per ton of CO_2 emissions. Currently, nuclear power is simply not likely to be competitive with coal and gas, and recent

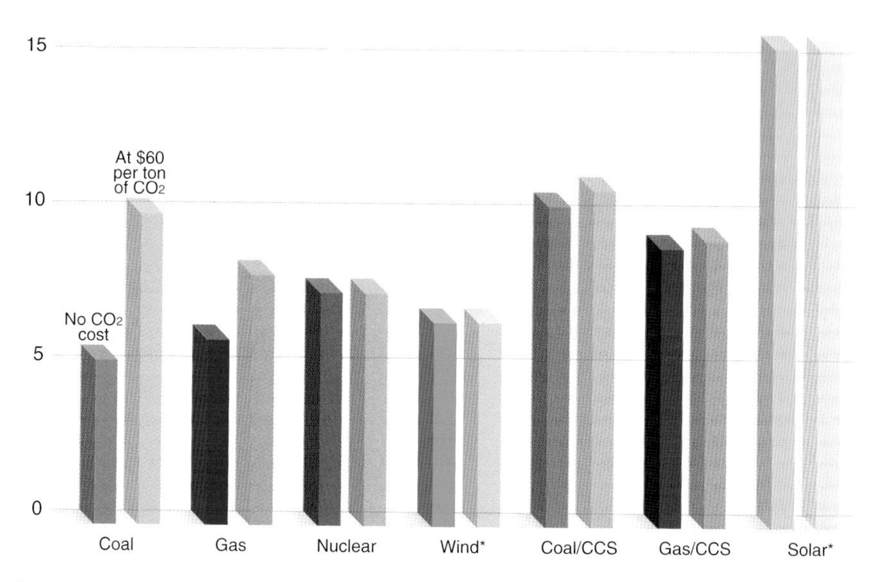

Figure 14.1 Average U.S. cost of electricity generation in 2025.
Source: ExxonMobil, the Outlook for Energy.

deployment of hydraulic fracturing technology opens vast potential supplies of natural gas, and hence low prices, for decades to come. But with a sufficiently high price of carbon emissions it could become competitive. With the long lead times to site, commission, and build additional nuclear power, including regulatory approvals, financing, and legal delays, reducing the uncertainty over any future pricing of carbon would facilitate planning and funding.

The United States currently has a plethora of alternative energy mandates, production, and consumption subsidies. These include federal and (especially in California) state tax breaks, loans and guarantees, utility purchasing requirements, ethanol mandates, etc. The result is far too much micromanagement and uncertainty. John Cogan and John Taylor estimate the federal subsidies alone for wind and solar at more per

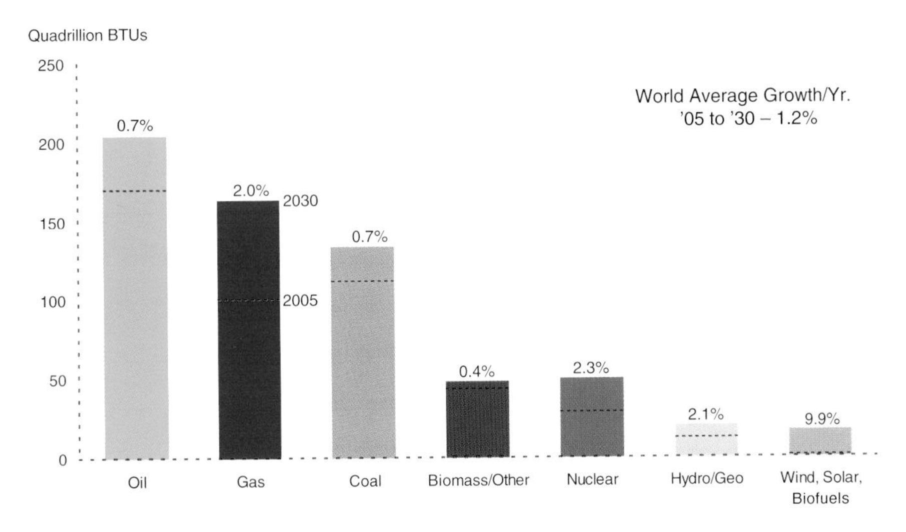

Figure 14.2 Energy Mix Continues to Evolve.
Source: ExxonMobil 2010 Energy Outlook.

kilowatt hour (kwh) than the current price of electricity. Replacing this morass with a revenue-neutral carbon tax would be a big improvement.

There are several detailed long-term (through 2030 or 2035) energy projections. The U.S. Energy Information Agency (EIA) publishes detailed U.S. data and forecasts, as do IHS Global Insight and Inforum. ExxonMobil produces detailed analyses of *global* energy supply and demand, by sector, fuel, and geography, which I cite below. There are several features relevant to nuclear power. Historical, current, and 2030 projected global energy supply and demand are presented in Figures 14.2 and 14.3. As can be seen, nuclear power is projected to grow substantially more rapidly than most other sources. To be sure, the resulting demand for transportation fuel, commercial power, and household electricity depends substantially on the global rate of economic growth and on the distribution of that growth among various countries with different energy mixes and intensity.

I discuss global totals to make the generic point, since some of the issues are global, e.g., oil is traded on a global market and the potential impact of greenhouse gas emissions on climate depends on global emissions. Electricity cannot easily be stored, so some energy issues relevant

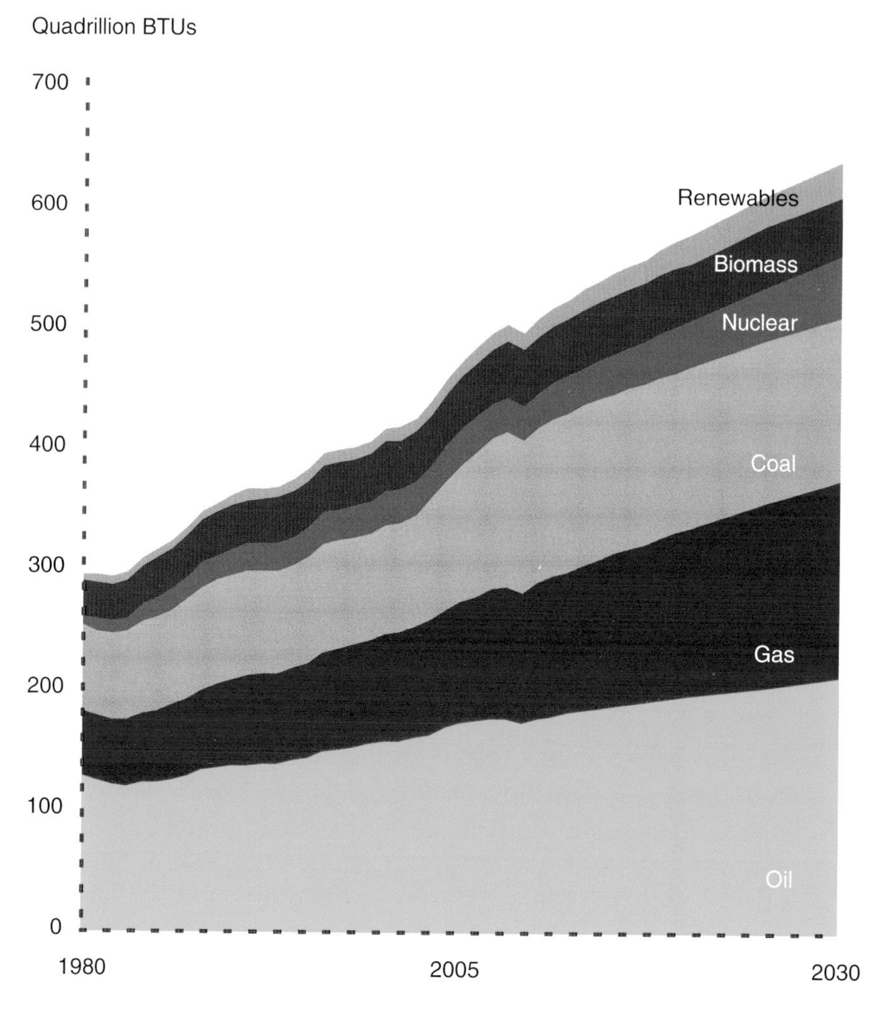

Figure 14.3 Global Demand by Fuel.
Source: ExxonMobil, the Outlook for Energy.

to nuclear power are local or regional. Nuclear power tends to be produced and transmitted on a regional basis (although some regions are quite large and span national borders). On the other hand, an expansion of nuclear power would cause substitution away from other primary fuels which are more widely and globally traded, such as natural gas and coal.

Table 14.1 on pages 303 and 304 outlines the basic data on recent energy supply and demand and ExxonMobil's projections to 2030, by

Table 14.1 ExxonMobil's *The Outlook for Energy*

Regions	Energy Demand (Quadrillion BTUs)							Average Annual Change			Share of Total	
	1980	1990	2000	2005	2010	2020	2030	1980-2005	2005-2030	2010-2030	2005	2030
World	**296**	**359**	**414**	**469**	**506**	**575**	**636**	**1.9%**	**1.2%**	**1.2%**	**100%**	**100%**
OECD	169	190	224	233	223	230	230	1.3%	0.0%	0.2%	50%	36%
Non OECD	127	170	190	237	282	346	406	2.5%	2.2%	1.8%	50%	64%
North America	87	95	114	116	111	113	113	1.2%	-0.1%	0.1%	25%	18%
United States	75	81	96	97	92	92	91	1.1%	-0.3%	0.0%	21%	14%
Latin America	13	15	20	22	26	32	39	2.2%	2.4%	2.1%	5%	6%
Europe	67	74	79	83	80	82	83	0.9%	0.0%	0.2%	18%	13%
European Union	63	68	72	76	73	73	73	0.7%	-0.1%	0.0%	16%	11%
Russia/Caspian	46	57	38	41	40	42	45	-0.5%	0.4%	0.6%	9%	7%
Africa	13	17	22	26	28	34	42	2.8%	2.0%	2.1%	5%	7%
Middle East	8	11	18	23	29	36	42	4.3%	2.5%	2.0%	5%	7%
Asia Pacific	63	91	124	159	193	235	273	3.8%	2.2%	1.7%	34%	43%
China	23	33	44	69	92	114	132	4.4%	2.7%	1.8%	15%	21%
India	8	13	19	22	28	35	45	3.9%	3.0%	2.4%	5%	7%
World Energy by Type Primary	**296**	**359**	**414**	**469**	**506**	**575**	**636**	**1.9%**	**1.2%**	**1.2%**	**100%**	**100%**
Oil	128	136	156	171	173	191	204	1.2%	0.7%	0.8%	36%	32%
Gas	54	72	89	101	112	138	164	2.5%	2.0%	1.9%	21%	26%
Coal	70	86	90	112	128	133	134	1.9%	0.7%	0.2%	24%	21%
Nuclear	7	21	27	29	28	38	50	5.6%	2.3%	2.9%	6%	8%
Biomass/Waste	29	36	41	44	47	48	48	1.6%	0.4%	0.1%	9%	8%
Hydro	6	7	9	10	11	14	16	2.2%	2.0%	2.0%	2%	3%
Other Renewables	0	1	3	3	7	13	20	8.0%	7.4%	5.8%	1%	3%
End-Use Sectors – World Industrial												
Total	**124**	**138**	**148**	**169**	**185**	**206**	**227**	**1.2%**	**1.2%**	**1.0%**	**100%**	**100%**
Oil	47	45	50	55	56	61	65	0.7%	0.7%	0.7%	33%	29%
Gas	28	30	38	40	43	51	60	1.5%	1.6%	1.7%	24%	27%
Coal	27	29	25	32	38	38	35	0.8%	0.4%	-0.4%	19%	16%
Electricity	14	18	21	25	30	38	47	2.5%	2.5%	2.3%	15%	21%
Other	9	15	14	16	18	19	19	2.1%	0.7%	0.3%	10%	8%

Table 14.1 ExxonMobil's *The Outlook for Energy* (*continued*)

Regions	Energy Demand (Quadrillion BTUs) 1980	1990	2000	2005	2010	2020	2030	Average Annual Change 1980-2005	2005-2030	2010-2030	Share of Total 2005	2030
Residential/Commercial												
Total	**71**	**87**	**98**	**107**	**111**	**124**	**134**	**1.6%**	**0.9%**	**1.0%**	**100%**	**100%**
Oil	14	13	16	16	15	15	15	0.7%	-0.2%	0.0%	15%	11%
Gas	13	17	21	22	24	28	32	2.1%	1.4%	1.4%	21%	24%
Biomass/Waste	23	26	29	31	32	31	29	1.3%	-0.3%	-0.5%	29%	22%
Electricity	10	16	23	27	30	39	49	4.0%	2.4%	2.5%	26%	37%
Other	11	15	9	10	10	10	9	-0.6%	-0.2%	-0.2%	9%	7%
Transportation												
Total	**53**	**65**	**80**	**90**	**96**	**112**	**124**	**2.2%**	**1.3%**	**1.3%**	**100%**	**100%**
Oil	51	64	79	88	92	106	115	2.2%	1.1%	1.1%	98%	93%
Other	2	1	1	2	4	6	9	1.3%	5.7%	3.8%	2%	7%
Power Generation – World Primary	**78**	**118**	**143**	**169**	**187**	**224**	**261**	**3.1%**	**1.7%**	**1.7%**	**100%**	**100%**
Oil	17	15	12	12	10	9	9	-1.3%	-1.2%	-0.6%	7%	3%
Gas	13	24	30	38	44	58	70	4.2%	2.5%	2.4%	22%	27%
Coal	34	48	61	76	87	93	97	3.3%	1.0%	0.6%	45%	37%
Nuclear	7	21	27	29	28	38	50	5.6%	2.3%	2.9%	17%	19%
Hydro	6	7	9	10	11	14	16	2.2%	2.0%	2.0%	6%	6%
Wind	0	0	0	0	1	4	7	44.2%	12.8%	9.3%	0%	3%
Other Renewables	1	3	4	5	6	8	11	7.1%	3.4%	3.1%	3%	4%
	Electricity Demand (Terawatt Hours)											
World	**7139**	**10136**	**13163**	**15657**	**17845**	**23061**	**28628**	**3.2%**	**2.4%**	**2.4%**	**100%**	**100%**
OECD	4948	6630	8559	9307	9352	10668	11744	2.6%	0.9%	1.1%	59%	41%
Non OECD	2192	3506	4604	6351	8493	12393	16884	4.3%	4.0%	3.5%	41%	59%
	Energy-Related CO_2 Emissions (Billion Metric Tons)											
World	**18.6**	**21.3**	**23.5**	**27.2**	**29.5**	**32.5**	**34.6**	**1.5%**	**1.0%**	**0.8%**	**100%**	**100%**
OECD	11.0	11.3	12.7	13.2	12.4	12.0	11.0	0.7%	-0.7%	-0.6%	49%	32%
Non OECD	7.6	10.0	10.7	14.0	17.2	20.6	23.6	2.5%	2.1%	1.6%	51%	68%

Rounding of data in the Outlook *may result in slight differences between totals and the sum of individual components.*

geography, fuel type, and sector. Given projected population and real GDP growth and modernization of developing economies, substantially more energy will be needed. New technology will be required to increase the efficiency of energy use, expand supply, and replace the sizable decline in many existing oil fields, as well as to mitigate emissions. The International Energy Association[4] estimates over $25 trillion will need to be invested over the period 2008–2030 to meet the world demand for energy.

Energy demand is projected to increase 26 percent by 2030 due to real per capita GDP growth, especially in geographies with current low energy use (1.4 billion people currently lack electricity). Energy use in the United States and European Union is projected to be flat (ExxonMobil's forecast is on the low side for the United States and European Union, compared to some others), whereas strong growth in Latin America, the Middle East, Africa, and Asia accounts for *all* the energy demand growth. The forecast is for a somewhat more diverse mix of affordable fuels, but the time frame is likely far too short for any radical transformation. Note these projections already assume that substantial, as yet unknown, technological improvements will occur to make major efficiency gains and alternatives to coal and oil more technically feasible and cost-competitive at scale. They already incorporate projections of more regulation, mandates, and subsidies along the lines we have witnessed in Europe and the United States in recent decades.

A more substantial change in primary energy type is eventually possible, assuming numerous more fundamental technical breakthroughs that dramatically reduce cost of alternatives to oil and coal, as demonstrated historically in Figure 14.4. Over two-plus centuries, global fuel use shifted from almost 100 percent biomass—wood and dung (still predominant in some poor parts of the world)—to primary reliance on coal in the first half of the twentieth century to heavy reliance on oil. Later came the growing importance of natural gas and finally hydropower, other renewables, and nuclear playing a small but growing role. Note that Figure 14.4 is in percentages of a rapidly growing energy use total. For example, total energy use is projected to double from roughly

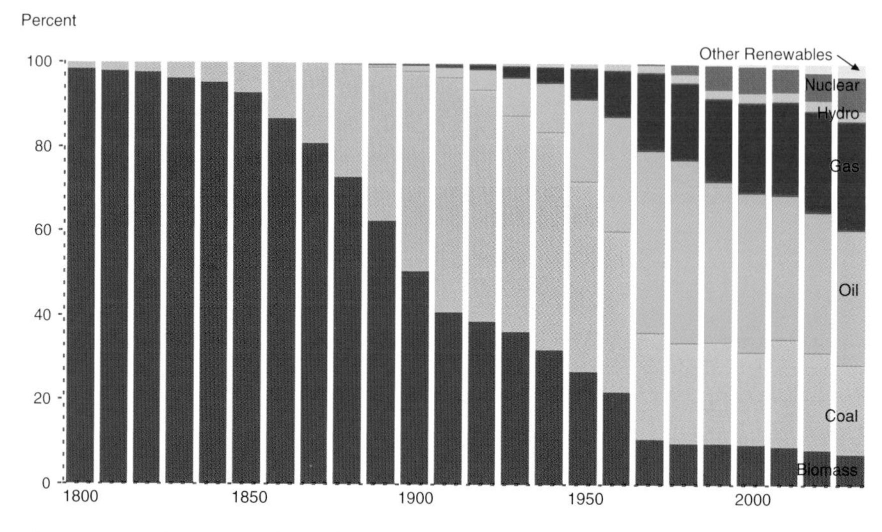

Figure 14.4 Energy Needs Evolve Over Time.
Source: Smil. *Energy Transitions;* ExxonMobil 2010 Energy Outlook.

300 quadrillion BTUs to 600 quadrillion in the half-century from 1980 to 2030.

Estimates of the likely growth of nuclear power use differ somewhat. Currently in the United States, the EIA estimates that 799 of 4,015 billion kilowatt hours of electricity (bkwh), or 20 percent, are derived from nuclear power. By 2035, the EIA estimates, that figure will grow to 874, an increase of 9.4 percent. With total electricity production growing to 5,181 bkwh, the share of nuclear power is projected to fall slightly to 17 percent. The corresponding U.S. projections for 2035 from IHS Inc. are 1,163 bkwh out of 6,025, a growth of about 15 percent and a share of 19 percent.

ExxonMobil's global forecast reflects the rapid growth in energy and nuclear power in the developing world. Nuclear is projected to grow from 28 quadrillion BTUs in 2010 to 50 quads in 2030, a global growth of 79 percent and a growth in the share of *total global* energy (not just electricity) from 6 percent to 8 percent. With nuclear, as with overall energy, the growth is projected to be outside the members of the Organisation for Economic Co-operation and Development (OECD).

The likely evolution of nuclear power in the United States will reflect the interaction of technological, economic, environmental, and political factors. Creation of a framework for long-run private investment decisions depends on prompt rationalizing of the regulatory, legal, tax, subsidy, and carbon pricing infrastructures. Also needed are sensible, predictable, permanent policies, with due regard to private and regulatory incentives. This might well lead to a more positive outlook for the growth of American nuclear power.

Notes

1. In February, 2012, the Nuclear Regulatory Commission approved the construction of two nuclear reactors at the Vogtle plant in Georgia.
2. "The Capture Theory of Regulation," chapter 11 in this volume.
3. "A Comparison of Government Regulation of Risk in the Financial Services and Nuclear Power Industries," chapter 13 in this volume.
4. International Energy Association, *World Energy Outlook*, November, 2009, pp. 104–107.

Session IV

Media and Public Policy

Media and Public Policy

15

JIM HOAGLAND

Journalists who swim in the mainstream often take comfort in seeing themselves as observers and not activists or advocates. But columnists and authors do not have such shelter available. We are paid to put opinions and perspective into newspapers, on the air, and into books. So George Shultz and Sid Drell have invited this group to analyze the media's contribution to the public discourse on nuclear risk with hopes to improve the discourse and public understanding of the opportunities and challenges surrounding the nuclear enterprise.

I am struck by an imbalance in the coverage of the two components of the nuclear enterprise. On the one hand, those of us who have directed media coverage have knowingly sought to influence broad public attitudes on the problems of weapons proliferation and, to a lesser extent, the safety (or lack thereof) of nuclear arsenals and materials. Major newspapers, magazines, and broadcast networks devote significant resources to developing their own expertise on national security and strategic affairs, as well as on foreign policy and global politics. We have made sure that our readers and viewers are exposed to the views and estimates of government and private authorities on the risks that nuclear arsenals and weapons proliferation pose to their well-being. It is perhaps telling that the harshest and most pervasive criticism leveled at the media in recent years centers on the accusations that we went too far in emphasizing the presumed threat posed by weapons of mass destruction that Iraq's

Saddam Hussein was widely assumed to possess. The most graphic depiction of the nuclear threat during the Cold War was cartoonist Herblock's stumble-bum, menacing A-Bomb that was part malignant and part stupid.

The investment by the media in security policy coverage has helped establish fairly clear, if shifting, contours for public opinion in Western democracies. Since the end of the Cold War, nuclear weapons have been seen by and large as necessary evils, at least when they are controlled by "responsible" countries. They are acknowledged evils, since those countries have pledged in the Non-Proliferation Treaty (NPT) to abolish them at some unknown future date. It is a sign of the times that the nuclear disarmament movement—which once produced emotional mass demonstrations and unpredictable political upheaval—is now headed up by responsible, sober authority figures such as George Shultz, Bill Perry, Sam Nunn, Henry Kissinger, and Sid Drell. The nuclear arsenals of the United States, Russia, Britain, France, and even China can now be analyzed in a relatively rational and orderly fashion in public discussion, with a clearer and more practical commitment to, and consensus on, dismantling them permanently beginning to emerge.

The focus of public anxiety today is the proliferation of nuclear weapons into the hands of less stable states or potentially of nuclear materials to terrorist gangs. There is a strong fear in the informed public that a tipping point is fast approaching as North Korea returns to a bellicose nuclear diplomacy, Iran augments its enrichment program despite international censure, and reports of a black market in nuclear materials periodically surface. This anxiety lies behind the widespread public support for diplomatic and other efforts to curb the programs of North Korea and Iran and to prevent a new arms race in the Middle East in which Arab states seek to counter not only Israel but also Iran in getting the bomb. And there is justifiable concern that a nuclear exchange between India and Pakistan comes closer as each nation continues to expand both its nuclear arsenal and its capacity for nuclear energy generation, activities bound together by the nuclear fuel cycle.

As David Hoffman's illuminating and forceful paper suggests (see chapter 16 in this volume), we have not done nearly as well in helping inform public opinion on nuclear energy. The media have not devoted the resources and effort that we put into covering proliferation, nuclear diplomacy, and stockpile safety into analyzing the nature of risk and benefit of nuclear energy production. Instead, as a group, journalists seem to share (and perhaps therefore help perpetuate) the fluctuating but usually deeply rooted ambivalence that is captured by public opinion polling in countries where nuclear energy is generated. The best known cultural icons of nuclear power on American television are, alas, Homer Simpson and his greedy employer, Mr. Burns.

With a few notable exceptions, media attention is not paid to the companies that produce nuclear energy, their safety standards and records, and the government agencies that regulate them—until disaster strikes. Even then the media interest tends to be episodic and fleeting.

Actually it is worse than that. When the media make serious efforts to investigate nuclear energy, they can come up short. A good example of this is the four-part series that the Associated Press distributed in June 2011. The stories elicited a sharp rejoinder from the industry, which can often be considered a sign of success by the journalists involved. But the *Columbia Journalism Review* also raked AP over the coals for overreaching and for failing to support many of its generalizations. Even worse, the series did not receive much attention from leading newspapers and broadcast networks. It seemed to prove the adages that no good deed goes unpunished and that we must indeed be careful what we wish for.

David's paper explores reasons behind the shortcomings and suggests six corrective measures that could help redress the imbalance.

From my readings, I think important findings on the disasters at Three Mile Island, Chernobyl, and Fukushima include the following.

- The interaction between the media and the authorities in these three cases added to public confusion rather than to clarity. Lessons developed out of the two earlier incidents were never learned or were forgotten before and during Fukushima. The poor communication among government and industry officials in all three cases led to briefings and statements to the media that at times stirred unnecessary alarm—but at least as often suppressed or minimized information that could have saved lives or eased panic and anxiety. (The information problems were compounded in the Soviet Union by deliberate deception on the part of government officials.)

- The experts and authorities lacked experience and discipline in communicating and explaining bad news in a transparent and credible way. Reporters and editors lacked experience and expertise in science that could have helped reduce or avoid crucial misunderstandings.
- Disasters and the coverage they inspire do not occur in a vacuum. Appreciation of risk is a social construct that includes not only current information but preexisting conclusions and biases. We still live under the shadow of the fear of radiation stirred by atmospheric testing six decades ago. Both authorities and journalists need to keep this in view as they attempt to communicate technical and other data that are susceptible to exaggerated reaction.
- The media have not been sufficiently clear with the public about what they do not know, and do not pursue, both between and during crises. Government and industry bodies appear to have done more over the long term to investigate and publicize the causes and possible remedies for communications and information failures than have the media. A greater openness by media organizations to participating in training exercises such as periodic disaster drills and seminars organized by public authorities and reactor operators would be useful.
- Experts can suffer a failure of imagination when it comes to considering catastrophic risk, especially when doing so would call into question the value and existence of the activities they have been asked to evaluate. In the run-up to our lingering financial crisis, computer models showed bankers what they needed to see: lucrative subprime mortgages and the synthetic credit devices based on them would never crash. At Fukushima, a hydrogen buildup which engineers did not think could happen became a crucial element in the disaster. Just as journalists sometimes talk about a story that is too good to check—for fear it will be knocked down—

catastrophic results threaten to be too bad to check. The media need to be more aggressive in their task of establishing accountability, in examining and challenging industry, regulators, and the government about what is not being anticipated.

- Perhaps most importantly for our discussion, it is clear that coverage of the three disasters had an immediate and dramatic impact on public attitudes toward nuclear power—in a strong, negative direction. In Japan, public *approval* of more nuclear plants stood at 82 percent six years ago, but swung to 70 percent *disapproval* after Fukushima. Germany permanently shut down eight of its seventeen reactors and said it would abandon nuclear energy by 2022. Coverage and commentary do matter.

It is true that, over time, the impact of Three Mile Island (no deaths) and Chernobyl (thirty-one immediate fatalities from blast and radiation exposure) faded as time went on and other issues, such as global warming, gained saliency. Support for increasing nuclear energy production moved higher in the United States in recent years as gasoline prices climbed. But Fukushima's three-tiered calamity is likely to have a more pervasive and durable effect—at least in the developed world, which is simultaneously experiencing a revolution in media affairs that adds pressures of urgency and over-simplification to an astounding availability of information and misinformation online. One of the questions we will want to consider is whether the vast Fukushima coverage has established a new pattern for the public in obtaining information about nuclear disasters. It is important to note, as Sharon M. Friedman did in the September 19, 2011, edition of the *Bulletin of the Atomic Scientists*, that although almost all of the casualties (estimated at nearly 16,000 deaths, with more than 3,000 people still missing) and great physical destruction were caused by the earthquake and tsunami, most of the coverage concentrated on the nuclear plant damage and concern about radiation.

There is an allied point to be made about the perception of risk, which for journalists exists primarily in the psychological realm, not the mathematical and economic ones most of you here deal in. There was little if any discussion in the mainstream media of the fact that the damage wrought by the reactor meltdowns in Japan represents a small fraction of the destruction that would have occurred if a nuclear weapon had exploded. That is in many ways the untold story out of Fukushima.

But we are here to explore that story. And in mentioning the AP series and the criticism from within the trade it elicited, I was pointing to the self-correcting mechanisms that are embedded in American journalism and in American society at large. There is a final cultural reference point we should keep in mind, and it is William Faulkner's magisterial Nobel Laureate speech devoted to the nuclear threat, and Faulkner's judgment that in the end mankind will prevail.

References

Associated Press four-part series on nuclear safety:
Part 1: http://www.seacoastonline.com/articles/20110620-NEWS-106200320
Part 2: http://topnews360.tmcnet.com/topics/associated-press/articles/2011/06/20/188209-bc-us-aging-nukes-part-2hfr-us.htm
Part 3: http://abcnews.go.com/US/wireStory?id=13936508#.T5Bex6sV2D8
Part 4: http://news.yahoo.com/ap-impact-nrc-industry-rewrite-nuke-history-070132412.html

Columbia Journalism Review critique of series: http://www.cjr.org/audit_arbiter/a_frustrating_ap_series_on_nuc.php?page=all

Industry Response to Series: http://safetyfirst.nei.org/safety-and-security/setting-the-record-straight-nei-responds-to-ap-series-on-nuclear-energy

Bulletin of the Atomic Scientists: http://www.thebulletin.org

The Nuclear Credibility Gap: Three Crises

16

DAVID E. HOFFMAN

Introduction

The most alarming nuclear accident in U.S. history began on March 28, 1979, at the Three Mile Island nuclear plant. Two days later, on the morning of March 30, the Nuclear Regulatory Commission (NRC) was debating whether to recommend that Governor Richard Thornburgh of Pennsylvania order an evacuation. The chairman of the commission, Joseph M. Hendrie, remarked that in the absence of reliable information from the plant, he and the governor were "like a couple of blind men staggering around making decisions." They struggled with a lack of data about the condition of the crippled reactor, poor communications with each other and the public, and a panicky population. While the accident caused no significant radioactive contamination outside the plant—nor any casualties—it became a potent symbol of public fears about the risks of nuclear power.[1]

The chaos and confusion were repeated during the Chernobyl nuclear disaster in 1986, and in Japan following the earthquake, tsunami, and multiple meltdowns of nuclear fuel at the Fukushima Daiichi power station on March 11, 2011. The design of the plant, the cause of the mishaps, and the political situation were different in each case. But there are important parallels that shed light on why public trust in nuclear power has ebbed.

Journalism played its own role in these crises. Today, in the aftermath of Fukushima, fresh doubts have erupted about civilian nuclear power, and some have pointed to flaws in the news media coverage—hype, ignorance and exaggeration—as a cause of the doubts. This paper argues that journalism about radiation and science can be improved, but the credibility gap runs far deeper. Nuclear power suffers from an enduring lack of public trust rooted in multiple factors, of which the news media are just one. Paul Slovic, a research psychologist well known for his work on risk perception, has written that "danger is real, but risk is socially constructed. Risk assessment is inherently subjective and represents a blending of science and judgment with important psychological, social, cultural, and political factors." He adds:

> One of the most fundamental qualities of trust has been known for ages. Trust is fragile. It is typically created rather slowly, but it can be destroyed in an instant—by a single mishap or mistake. Thus, once trust is lost, it may take a long time to rebuild it to its former state.[2]

When public trust is broken today, the impact is instant and global. News has become a vast, transnational force—one that never stops at passport control. As Fukushima has shown, a failure in one country can undermine public confidence in many others, even in those without the same shortcomings in their nuclear power industries. The news media can bring immense, transformational influence on how people think about danger and how they perceive risk. After Fukushima, public opinion surveys showed a sharp rise in perceptions of risk in nuclear energy. Three Mile Island and Chernobyl were followed by major changes in nuclear plant design and training, but many of the worst mistakes in communication were repeated at Fukushima. What can be done differently next time?

Three Crises

Three Mile Island

Peter M. Sandman, a professor and consultant on risk communications, has written, "What went wrong at TMI—really, really wrong? The communication."[3] Especially in the first three days, the media and the public were given widely divergent messages about the severity of the accident and the risks posed by radiation from the damaged reactor. The presidential commission which investigated the accident concluded that neither the utility nor the NRC had specific plans in place for providing information to the public and the news media. The commission found that "during the accident, official sources of information were often confused or ignorant of the facts. News media coverage often reflected this confusion and ignorance."[4]

On March 28, 1979, at 4 a.m., a valve stuck open in Unit 2 at the Three Mile Island plant, a pressurized water reactor running at 97 percent of capacity. The valve permitted cooling water to escape. Then, operators made an error. Thinking that the system was filling up with water when in fact it was losing water, they shut down some emergency pumps. The operators were disoriented by the complexity of the plant and a large number of alarms; more than 100 went off simultaneously. The plant began to undergo loss of coolant, but the operators did not recognize this. The indicator about the stuck valve was located out of sight, and remained unnoticed.[5]

The reactor immediately shut down. The control rods moved into place to end the fission reaction. But, when a reactor shuts down, there is a latent amount of energy, known as decay heat. In a reactor that has been running for a while, the decay heat is about 7 percent of the original power level at the moment of shutdown. Removing decay heat is a concern for reactor shutdowns and especially for reactor loss-of-cooling accidents.

At Three Mile Island, the cooling of the core was inadequate. Water turned to steam. The cladding on the fuel rods ruptured and, reacting with the steam, produced large amounts of hydrogen. Upper sections of the core slumped into a molten mass. Researchers later discovered that about half the core had melted during the early stages of the accident. Later investigations estimated that at some parts of the core, the temperature reached 4,000 degrees F.[6]

The confusion at the plant led to opaque and contradictory public statements. The utility, Metropolitan Edison, at first "offered bland affirmations about the safety of the plant that became increasingly less credible," according to a history of the accident sponsored by the NRC. Then came a statement that the plant had a malfunction and would be "out of service for about a week." The next statement, at 8:30 a.m., said the plant had been "shut down due to a mechanical malfunction," but "there have been no recordings of significant levels of radiation and none are expected outside the plant." In fact, extremely high levels of radiation had been recorded *inside* the plant. The utility's superintendent for technical support, George Kunder, who had arrived at the plant at 4:50 a.m. to help the besieged operators, had exclaimed upon hearing the radiation levels, "Oh my God, we're failing fuel." This was not what the public was told. According to the official history of the events:

> Because of the uncertainty that prevailed at the plant, the information that the utility provided to government agencies on March 28 was usually fragmentary and sometimes contradictory or ambiguous. The reports it issued to the press and public understated the severity of the accident. In turn, state government and NRC officials all too frequently circulated confusing or erroneous information about the accident.[7]

Lt. Gov. William W. Scranton III gave a press conference that morning in which he said "everything was under control," but his answers were "confusing about the threat of radioactive releases."[8]

When officials of the Nuclear Regulatory Commission attempted to compose a press release in the afternoon of March 28, they fretted about using the word "accident," according to transcripts of their discussions. In the draft press release, the word was used twice. Commissioner Richard T. Kennedy feared this would imply the plant was heading for a meltdown, as depicted in the then popular movie, *The China Syndrome*. "Is this an accident?" he asked. Eventually, the commissioners decided to remove the second mention of the word *accident*. But the plant had undergone a partial meltdown.

On March 30, operators deliberately vented a buildup of radioactive gas by opening a valve for an extended period. A utility company helicopter, 130 feet directly above, took a reading of 1,200 millirems per hour. This information was passed to the state and to regulators, but was garbled—wrongly received as a reading on the ground and off-site, suggesting a spread of radiation. This led Thornburgh to call for evacuation of children and pregnant women within five miles. The off-site readings were, in fact, between fourteen and twenty-five millirems per hour, and dropping, which was not as alarming. "As a result of poor communications and enormous misunderstandings, the 1,200-millirem reading set off a crisis," the official history concluded.

The presidential commission which investigated the accident said that at this point, two days after the accident, the senior management of the NRC "could not and did not develop a clear understanding of conditions at the site." There was confusion about whether a hydrogen bubble that had built up inside the reactor could explode.[9] On March 31, Hendrie suggested to the press that it might explode, and that a wider evacuation might be necessary. This set off a near-panic in Pennsylvania. An Associated Press report about a possible hydrogen explosion, based on Hendrie's remarks and other reporting, led thousands of people to flee their homes.[10] Only late Saturday night did the NRC spokesman, Harold Denton, attempt a correction, saying that there was no danger of a hydrogen explosion. But the misleading information was already widely disseminated. The confusion continued on April 1, when President Jimmy Carter visited the site. New measurements showed the bubble was shrinking, but some on the NRC staff were still uncertain that this is what was happening inside the reactor.

The presidential commission later concluded:

> The NRC staff was confronted with problems it had never analyzed before and for which it had no immediate solutions. One result of these conditions was the computational errors concerning the hydrogen bubble, which caused the NRC to misunderstand the true conditions in the reactor for nearly 3 days.[11]

> By late Sunday afternoon, NRC—which was responsible for the concern that the bubble might explode—knew there was no danger of a blast and that the bubble appeared to be diminishing. It was good news, but good news unshared with the public. Throughout Sunday, the NRC made no announcement that it had erred in its calculations or that no threat of an explosion existed.[12]

The presidential commission noted that "the response to the emergency" at Three Mile Island "was dominated by an atmosphere of almost total confusion. There was a lack of communication at all levels."

Chernobyl

First, it must be noted that the Soviet Union was a closed system, without channels to challenge the authorities. The Chernobyl disaster must be seen as part of that system—indeed, as a bell tolling for its demise. Such closed systems exist today; China may be the largest example. The Chernobyl experience is relevant, even if the Soviet Union is gone.

In particular, the Soviet media were tools of the power structure. There was no way for them to convey to the public the multiple deceptions surrounding Chernobyl events. This was illustrated by the experience of one editor at Pravda, who visited the scene with a group of journalists. The editor wrote a cautious report for the newspaper, but privately delivered a damning, highly critical report to the Kremlin.[13]

The accident began Saturday, April 26, 1986, during an operator test at Reactor No. 4. The reactor was a mammoth block of graphite honeycombed by fuel rods and cooled by water pumps. Unlike reactors in the West, the Soviet design lacked an overarching containment structure.[14]

Chernobyl was put into operation with a known flaw. If outside electric power were suddenly cut off, forty seconds would be needed to start auxiliary diesel engines to power generators and keep water pumps moving. Without power for this time, the pumps could not force water through the reactor. Plant designers were still trying to fix this forty-second gap in 1986. On the night of April 26, an improvised work-around was being tested. The operators' objective was to determine whether a generator could be hitched up to the still spinning (but slowing) turbine blades, and whether this would produce enough power to keep the pumps working for forty seconds.

After midnight Saturday, the reactor was powered down to very low levels for the test. Then, apparently because power was too low, the operators tried to power it up again, perhaps too quickly. Nuclear fission creates byproducts that must be allowed to dissipate before the reactor is powered up again, but this danger was ignored. As the operators powered up the reactor, a chain reaction began to spin out of control.

The operators hit the red panic button. They desperately tried to lower the control rods to stop the fission, but the rods, by some accounts, got stuck. As the heat inside the core skyrocketed, more of the water then turned to steam, which caused the core to get even hotter. At 1:23 a.m., two explosions rocked Chernobyl. These were extremely powerful, as the ongoing chain reaction continued to generate huge amounts of heat and pressure. The reactor blew apart, and the explosions were followed by fire. Radioactive materials—gases, graphite, and bits of broken fuel rods—were flung into the atmosphere. Some debris fell down near the site. Winds carried radioisotopes north and west across Europe. The initial explosion and contamination was one nightmare. Then came another: the graphite stack was on fire and burned for ten days, spewing more dangerous materials into the air.

While the graphite stack burned, Deputy Energy Minister Alexei Makukhin sent a report to Moscow saying that at 1:21 a.m. on April 26 an explosion occurred in the upper part of the reactor, causing fire damage and destroying part of the roof. "At 3:30, the fire was extinguished." Personnel at the plant were taking "measures to cool the active zone of the reactor." No evacuation of the population was necessary, the report said.[15] Almost everything in Makukhin's report was wrong. The reactor was still burning and was not being cooled, and the surrounding population should have been evacuated immediately. What the report omitted was even worse: at the scene, radiation detectors failed, firefighters and others were sent in without adequate protection, and officials were debating—but not deciding—about evacuation.

Soviet President Mikhail Gorbachev recalled many years later that he first heard of the disaster in a phone call at 5 a.m., but did not learn until the evening of April 26 that the reactor had actually exploded and there had been a huge discharge into the atmosphere. "Nobody had any idea that we were facing a major nuclear disaster," he recalled. "Quite simply, in the beginning even the top experts did not realize the gravity of the situation." Although Gorbachev does not say so explicitly, the reason for the lack of information was the Soviet system itself, which reflexively buried the truth. At each level of authority, lies were passed up and down the chain and the population was left in the dark. This distinguishes Chernobyl from the other two accidents which occurred in democracies with relatively open channels of information and free press.

By April 27, radiation was detected in Sweden at the Forsmark nuclear power station, one hundred miles north of Stockholm. The Swedes confronted the Soviets at midday Monday, April 28. A Soviet government announcement, under consideration by the Politburo, was delayed. Gorbachev later claimed there were two reasons for the delay: he lacked information and didn't want to create panic. The Kremlin eventually instructed the news media to distribute a statement so terse as to convey nothing of the catastrophic nature of the event. The announcement was issued at 9 p.m. on April 28:

> An accident has occurred at the Chernobyl Nuclear Power Plant, damaging one of the reactors. Measures are being taken to eliminate the consequences of the accident. The injured are receiving aid. A government commission has been set up.

On the next day, April 29, the Politburo decided to issue another public statement, saying the accident had destroyed part of the reactor building and the reactor itself and had caused a degree of leakage of radioactive substances. Two people had died, the statement said, and "at the present time, the radiation situation at the power station and the vicinity has been stabilized."

Throughout the crisis, the bosses of the Soviet state obfuscated. One of the first actions of the plant director was to cut nonessential telephone lines around Chernobyl.[16] An evacuation of the nearby town of Pripyat was begun thirty-six hours after the explosion; the second stage of the evacuation, a wider zone that eventually displaced 116,000 people, did not begin until May 5. The Communist Party in Ukraine insisted that May Day parades should carry on as usual in Kiev even though winds were blowing in that direction.

Thirty-one people were killed as direct casualties of the Chernobyl accident. By one estimate, up to 4,000 additional cancers may have resulted among the 600,000 people exposed to higher levels of radiation, such as liquidators (clean-up workers), evacuees, and residents of the most contaminated areas.[17]

In the aftermath, the Soviet Union and Russia carried out a modernization of nuclear power plants similar to Chernobyl, many of which still operate, and there was a major push to address nuclear safety around the world. Gorbachev was motivated to intensify his efforts to reduce nuclear weapons after what he had been through at Chernobyl. The Reykjavik summit meeting between Gorbachev and President Ronald Reagan followed within six months.

Fukushima

Japan had invested heavily in nuclear energy, which supplied about 30 percent of its overall power needs. Over decades, Japan's nuclear establishment devoted resources and effort to persuade the public of the safety and necessity of nuclear power. There were public relations campaigns, even government-mandated school textbooks with friendly views of nuclear power. Many Japanese had thus come to accept the conclusion that Japan's plants were absolutely safe. The earlier nuclear accidents at Three Mile Island and Chernobyl barely registered.[18]

Unlike the other two cases, Japan's nuclear accident began with a cataclysmic natural disaster. An offshore earthquake, magnitude 9.0, the largest in Japan's modern history, occurred at 2:46 p.m. on March 11, 2011, triggering a massive tsunami. Almost 16,000 people were eventually reported dead, with another 3,000 missing. The quake toppled power lines and separated the six-unit Fukushima Daiichi Power Station from the electricity grid. Three units were in operation at the time. Forty-one minutes after the quake, a tsunami wave hit the nuclear plant, followed seven minutes later by another.

The quake knocked out electric supplies to the power station. The reactors automatically inserted the control rods to shut down the fission reaction. Diesel generators activated to keep the water pumps working, as planned. However, within an hour, the tsunami inundated the diesel generators and key circuit boards. They stopped functioning. Without electricity, coolant water was not circulated by pumps, and boiled off. Although the control rods had been inserted, there was still decay heat in the reactor cores. The fuel rods in the cores of Units 1, 2 and 3 were exposed, leading to core meltdowns.

The zircaloy cladding of the rods oxidized, giving off hydrogen—the same problem that had been a source of worry at Three Mile Island. An explosion, probably caused by hydrogen, ripped through the building of Unit 1 on March 12 at 3:36 p.m. Two days later, another hydrogen blast tore through the building of Unit 3 at 11:01 a.m. And on March 15, an explosion occurred at Unit 4, where a large amount of spent fuel was in storage. The explosions destroyed the operation floor of each unit, making recovery even more difficult.

Japan's faith in nuclear power has been badly shaken by the disclosure, after the disaster, of flaws in plant design and lapses in management of the crisis.

The most serious flaw was contained in assumptions about a possible tsunami. The height of the one that occurred March 11 greatly exceeded worst-case levels assumed by designers. The plant's license stated the maximum tsunami height would be 3.1 meters. A separate Japanese civil engineering standard for nuclear facilities was 5.7 meters. But this tsunami reached 14 to 15 meters. The diesel generators and electric switches were located in a low-lying turbine building, vulnerable to failure when the waves rolled in. This error, had it been recognized and corrected, might have prevented the worst consequences of the accident.

The plants had no method to cope with a hydrogen buildup. The designers apparently never thought one would occur. "It was not assumed that an explosion in reactor buildings would be caused by hydrogen leakage," the government told the International Atomic Energy Agency (IAEA) in June. "As a matter of course, hydrogen measures were not taken."[19]

On March 27, the Tokyo Electric Power Company (TEPCO), which operated the plant, announced that readings in the water from the basement of the turbine building attached to Unit 2 contained 10 million times the normal amount of the radionuclide iodine^{134}, giving rise to fears that radioactive materials had breached the primary containment vessel and flowed into the turbine building. But later that day, the utility retracted the reading as erroneous.

Japan's Nuclear and Industrial Safety Agency (NISA) announced March 18 that the accident was a Level 5 event on the International Nuclear and Radiological Event Scale. This scale, devised by the IAEA, is supposed to "facilitate communication and understanding between the technical community, the media and the public on the safety significance of events. The aim is to keep the public, as well as nuclear authorities, accurately informed on the occurrence and potential consequences of reported events."

Level 5 is an "accident with wide consequences," in which there has been released "a quantity of radioactivity . . . equivalent to a release to the atmosphere of the order of hundreds to thousands of terabecquerels" of iodine^{131}. A Level 5 accident is one in which "some protective action" will probably be required, such as localized sheltering or evacuation. Nearly a month later, on April 12, Japan abruptly raised the alert to Level 7, which is defined as a release to the atmosphere of more than several tens of thousands of terabecquerels of iodine^{131} and involving health effects over a wide area. (For example, Chernobyl was a Level 7 event.) The higher level for Fukushima was based on estimates by two agencies in

Japan of releases at 370,000 TBq (NISA) and 630,000 TBq (the Nuclear Safety Commission). In a report to the IAEA, Japan acknowledged that the Level 7 alert "should have been made more promptly."[20]

Early in the disaster, Prime Minister Naoto Kan, a one-time grassroots activist who had long been suspicious of the close ties between industry and bureaucrats, challenged the leaders of TEPCO, saying they were withholding information. At one point, while the prime minister's office was debating the risks of injecting seawater to cool the reactor cores, they did not know that the injection was already under way. At another point, TEPCO ordered the plant director to stop the seawater injection, but he ignored the order.

Kan had left the handling of the crisis to the utility in the first three days, as he sought to cope with relief efforts for those left homeless by the earthquake. But after the second hydrogen explosion, he flew into a rage when the TEPCO president asked for permission to withdraw his staff from the crippled plant, a drastic step which would have implied abandoning control of three overheating cores and seven spent-fuel ponds. On March 15, Kan marched into TEPCO headquarters and installed one of his top aides to keep tabs on the company.[21] Meanwhile, the company president disappeared, suffering from illness.[22]

At the same time, distrustful of bureaucrats, Kan bypassed a formal emergency management system that had been set up in earlier years, relying instead on a tight circle of aides. As a result, the prime minister and his circle did not know for five days about the existence of a system of nationwide radiation detectors which could have provided better data for evacuation decisions.[23] The data from the system was withheld from the public for weeks. According to the *New York Times*, some people were evacuated directly into an area where winds were carrying radioactive materials because this information had been withheld.[24] Kan resigned in late August 2011, in large part because of public disappointment over his handling of the crisis.

Public Opinion and Building Trust

There's a temptation to say after an event like Fukushima that the public is ignorant and irrational: *If only they had the facts, they would understand.* By this rationale, what's needed is simply to provide more factual information about nuclear power, and public opinion will improve. While such an assumption is logical, there are other factors. Public opinion is also profoundly influenced by subjective feelings such as fear, the unknown, and loss of control. These must also be taken into account when considering attitudes in the future, and the news media play a very large role in how the public shapes such fears.

In general, the American public holds complex sentiments about nuclear power, which have fluctuated considerably over the last three decades.

Three years ago—when gasoline prices reached record highs—57 percent of Americans questioned said they approved of building new nuclear power plants to generate electricity, according to the *New York Times/CBS News* Poll. After Fukushima, 43 percent approved, while 50 percent did not. Back when this question was first asked in 1977, before Three Mile Island, support was as high as 69 percent; it dropped as low as 34 percent after the Chernobyl disaster in 1986.

Americans are even less likely to approve of building a new nuclear power plant in their community—six in ten disapprove. Americans have solidly opposed building new nuclear power plants in their communities since 1979.

Table 16.1 Building More Nuclear Power Plants

	3/11	7/08	4/07	6/01	6/91	5/86	4/79	7/77
Approve	43%	57%	45%	51%	41%	34%	46%	69%
Disapprove	50	34	47	42	48	59	41	21

Source: "Nuclear Power Loses Support in New Poll," *The New York Times*, March 22, 2011, http://www.nytimes.com/2011/03/23/us/23poll.html.

Table 16.2 Building More Nuclear Power Plants in Your Community

	3/11	4/07	6/01	5/86	4/79
Approve	35%	36%	40%	25%	38%
Disapprove	62	59	55	70	56

Source: "Nuclear Power Loses Support in New Poll," *The New York Times*, March 22, 2011, http://www.nytimes.com/2011/03/23/us/23poll.html.

However, when the question is asked in the context of global warming, support for nuclear energy is stronger. A June 2007 poll by the *Los Angeles Times* and Bloomberg found that if nuclear power is described as a way to reduce global warming, 56 percent of those asked said they would support building new power plants, compared to 34 percent against.[25]

Around the world, there are similar sentiments. In Japan, a 2005 poll found that 82 percent of those questioned favored building more nuclear plants or maintaining existing ones. But after Fukushima, polls showed that seven in ten Japanese favor scaling back or closing the country's nuclear power plants. Then Prime Minister Kan responded to the new sentiment by ordering safety checks on Japan's remaining reactors. Following the earthquake, worries over safety have forced TEPCO and other plant operators to keep more reactors idle after such checks. As of April 12, 2012, only one of Japan's 54 nuclear reactors was operating, although there were discussions about starting two others following safety checks. Without restarts, Japan will face a serious power shortage.[26]

Even in France, where nuclear plants supply three-quarters of the power, a poll in April showed more than half those questioned wanted to do away with nuclear power. In Germany, the government has finalized a package of bills that will phase out nuclear power plants, which generated 23 percent of the country's total energy use last year.

The data above suggest that public opinions of nuclear power have changed and will evolve further. But nuclear power is surrounded by enduring suspicions and intense emotions. Slovic has found that extremely negative attitudes and perceptions of the risks assumed to stem from nuclear energy stand in sharp contrast with technical assessments of nuclear energy's safety record. "Nuclear power has a special distinction in the perception literature," he found. "It is, to date, the technology hazard with the most negative and most problematic constellation of traits. It stands apart in having qualities that make it fearsome and hard to manage socially and politically."

Slovic's research found that when people were asked to evaluate the risks of different types of radiation, they saw medical x-rays as rather benign, but "nuclear power had the dubious distinction of scoring at or near the extreme negative end for most of the characteristics. Its risks were seen as involuntary, unknown to those exposed or to science, uncontrollable, unfamiliar, catastrophic, severe (fatal) and dreaded."[27]

He observed:

> The nature of any low-probability/high consequence threat is such that adverse events appear to demonstrate the dangerousness of nuclear technology but demonstrations of safety require a very long time, free of damaging incidents or incidents perceived as damaging.

Slovic added, "The problem is not due to public ignorance or irrationality, but is deeply rooted in individual psychology and in the adversarial nature of our social, legal and political systems of risk management." A wide range of factors come into play when people are making judgments about nuclear power—not only the history of these three accidents but also controversy over building new plants, over regulating them, over nuclear waste, and other issues. The prominent arguments between the nuclear industry, anti-nuclear groups, and other energy industries, as well as local conflicts over power plants, feed into risk perceptions. As Slovic noted in his 1999 paper, "Our social and democratic institutions, remarkable as they are in many respects, breed distrust in the risk arena."[28]

The Role of Journalism

At the time of Three Mile Island, there was no Internet or cellular phone communications. Most people did not have computers in their homes. The news was transmitted by radio, three major television networks, two news wire services, and several major newspapers, as well as many local ones. CNN was founded in 1980, the year after the accident. Reporting for television was usually not live, but filmed earlier for evening broadcast. Reporters on the scene did not have access to computers and high-speed communications to get information, nor could they verify facts as rapidly as they can today.

In covering the accident, the news media were caught up in the vortex of confusion. The Associated Press story about the possible hydrogen explosion at Three Mile Island was transmitted as an urgent advisory at 8:32 p.m. on Saturday, March 31. When the full story went out at 9:02, "the news it contained had already set off a furor." Radio and television stations broadcast the advisory even before seeing the full story. "Many people who had not evacuated the previous day did so on Saturday evening and Sunday in response to the hydrogen bubble scare," according to the history sponsored by the NRC. A third of those who fled did so at this time.

The hydrogen bubble scare, which originated with a press briefing by NRC Chairman Hendrie, points to a defining characteristic of the news media, then and now: in a fast-moving crisis, journalists are largely conveying information, not discovering it on their own. Lapses and errors made by sources are transmitted on the fly, and often amplified. This was the case with the bubble at Three Mile Island; reports about the possibility of an explosion were often confused with a nuclear bomb going off. "Hydrogen Blast Threat Looms," headlined the *New York Daily News* on Sunday. A Harris survey showed that after the crisis, 66 percent of those questioned believed that a nuclear power plant that failed could cause a "massive nuclear explosion," compared to 39 percent when the question was asked four years earlier.[29]

The first reporters to respond to a big, breaking story like Three Mile Island are often not trained in science or engineering. Lack of expertise remains an enormous weakness of the news media. Those organizations that can afford to employ specialists often have few, and there are fewer today than a generation ago because of steep economic and competitive pressures. The presidential commission on Three Mile Island found, "The reporters who covered the accident had widely divergent skills and backgrounds. Many had no scientific background."

The panel added:

> Because too few technical briefers were supplied by NRC and the utility, and because many reporters were unfamiliar with the technology and the limits of scientific knowledge, they had difficulty understanding fully the information that was given to them. In turn, the news media had difficulty presenting this information to the public in a form that would be understandable. This difficulty was particularly acute in the reporting of information on radiation releases. They also experienced difficulty interpreting language expressing the probability of such events as a meltdown or a hydrogen explosion; this was made even more difficult when the sources of information were themselves uncertain about the probabilities.[30]

Furthermore, the panel was quite blunt about how the news media and its sources functioned:

> Another severe problem was that even personnel representing the major national news media often did not have sufficient scientific and engineering background to understand thoroughly what they heard, and did not have available to them people to explain the information. This problem was most serious in the reporting of the various releases of radiation and the explanation of the severity (or lack of severity) of these releases. Many of the stories were so garbled as to make them useless as a source of information.
>
> We therefore conclude that, while the extent of the coverage was justified, a combination of confusion and weakness in the sources of information and lack of understanding on the part of the media resulted in the public being poorly served.
>
> In considering the handling of information during the nuclear accident, it is vitally important to remember the fear with respect to nuclear energy that exists in many human beings. The first application of nuclear energy was to atomic bombs which destroyed two major Japanese cities. The fear of radiation has been with us ever since and is made worse by the fact that, unlike floods or tornadoes, we can neither hear nor see nor smell radiation. Therefore, utilities engaged in the operation of nuclear power plants, and news media that may cover a possible nuclear accident, must make extraordinary preparation for the accurate and sensitive handling of information.[31]

In the case of Chernobyl, Soviet leaders strictly controlled the press and hid the truth of the accident from people for many days. But the overseas media were not constrained, and the reports of radiation spreading across Europe soon demonstrated to Gorbachev that, in a globalized world, he could not hide the consequences.

After the disaster, Gorbachev's emphasis on *glasnost*, or openness, grew significantly. At a July 3, 1986, Politburo meeting, he declared, "Under no conditions will we hide the truth from the public, either in explaining the causes of the accident nor in dealing with practical issues." He added, "We cannot be dodging the answers. Keeping things secret would hurt ourselves. Being open is a huge gain for us." Foreign Minister Eduard Shevardnadze's assistant Sergei Tarasenko said Gorbachev and Shevardnadze were shamed by the way the radioactive cloud floating over Europe had revealed what they failed to announce. "For the first time, they understood that you cannot cover up anything," Tarasenko said. "You can say, 'Nothing happened there,' but with radiation you cannot hide it. It will go in the air, and anyone will know it is there." Shevardnadze wrote in his memoirs that Chernobyl "tore the blindfold from our eyes and persuaded us that politics and morals could not diverge."[32]

Three decades after Three Mile Island, the news media have undergone a massive transformation. In the digital age, all reporters now have access to the Internet, even in remote locations, and the challenge becomes how to sift and analyze torrents of information. Mainstream press outlets are now competing with large and potent armies of online essayists and bloggers. The competition for clicks and viewers, especially online and in broadcast, has pushed everyone toward rushed and sometimes reckless journalism. There can be no pause, no void. Sensationalism is tolerated in order to win over larger audiences. A nuclear expert recalled his appearances on cable television during the Fukushima crisis: "Twenty-year-old producers making up questions—all sensational and leaving it to me to bring some balance and reality after I undid the false premise."[33]

At the same time, eyewitness reporting has been transformed as thousands of people on the scene can report and share their observations instantly by the combination of mobile devices and micro-blogging services like Twitter. In the days after the Fukushima accident, anyone using Twitter could read 140-character messages about the situation racing across the screen at a rate of one per second and continuing at this pace for hours at a time. The quality was both very high, pointing to original sources and breaking developments, and very low, filled with emotional over-reactions and exaggerations. On March 12, 2011, the day after the quake, tsunami, and nuclear accident, Twitter added 572,000 new accounts—more than the total print circulation of the *Philadelphia Inquirer* during the 1979 TMI crisis.

In Japan, the author found an amateur Web entrepreneur, Katz Ueno, who normally provided information about tourism and entertainment to an English-speaking audience via a Web site he founded known as *YokosoNews*. Ueno was one of the first to use Twitter to broadcast information about the events of March 11—his initial tweet carrying a tsunami warning was six minutes after the quake and he went live with his broadcasts forty minutes after the quake. Using a live Webcam at home, he translated to English the information broadcast in Japanese by state television NHK. This feed, which ran ten to eighteen hours a day for fourteen days, generated 1 million visitors to his Web site. It was not ideal—sometimes the information was fragmentary or opaque—but it was extraordinarily urgent and immediate.

This revolution has made crisis management even more difficult and more important. But some things have not changed since 1979. Fukushima unleashed a deluge of uninformed, exaggerated, and often alarming news reports, much of it circulated without restraint in cyberspace. The finding of radiation levels at 10 million times normal was widely broadcast, then retracted. The three hydrogen explosions were shown repeatedly on television—quite dramatic video of reactor buildings blowing up—without much context or additional information about why they had occurred. The very sight of a nuclear power plant exploding with smoke and a boom was enough to sow deep fears and confusion.

The people who lived around the Fukushima plant struggled to comprehend why their own government had ordered an eighteen-mile voluntary evacuation zone while the United States advised its citizens to evacuate an area fifty miles around the plant. (U.S. officials at the time were convinced that Japan was understating the damage at the plant, based on readings the United States made from unmanned aircraft and satellites normally used to monitor North Korean nuclear tests.)[34]

It is too early to render a verdict on the daunting enterprise of covering an accident like this. Much of the journalism was sound, skeptical, and commendable, but much wasn't.

Thomas Inglesby and Anita Cicero of the Center for Biosecurity at the University of Pittsburgh Medical Center said the Fukushima accident showed a lack of understanding in the media and public about the different types of radiation, measurement methods, dosage measurements, and health impacts, making it "difficult to explain . . . how radiation travels, how far it travels, and the degrees of harm caused by exposure in the short and long term."[35]

"Communications about the radiation risk in Japan have been poorly done," Slovic said. He asserts that what is needed is not only better information, but an appreciation that risk perceptions are driven by subjective values and assumptions. "The bottom line is that we need both emotion and reason," he said. The human mind has learned to assess risk through both facts and gut feelings, a sense of dread, he added. "There is a wisdom to dread that we should not casually dismiss."[36]

David Ropeik, a former environmental reporter for the *Boston Globe* who now lectures about risk communications, wrote in *Scientific American* that "the peril is not just the radiation. It's people's fears of radiation. Whether those fears are consistent with the evidence of the actual physical risk (they aren't) doesn't matter. Fear is real, and does real harm."[37]

The evident confusion and chaos of Fukushima suggest that problems first seen at Three Mile Island and Chernobyl still exist. Errors by government officials or utilities will be amplified, not corrected, in the coverage. Journalists often lack the basic skills for interpreting complex topics such as radiation science or nuclear reactor design. And public mistrust is deepened when governments are found to have hidden or misrepresented dangers. The outcome is already being seen in Japan, where the public has turned decisively against nuclear power after years of strongly supporting it.

The Crisis Next Time

Journalism alone cannot overcome the nuclear credibility gap. It is beyond the scope of this paper to prescribe how the industry and regulators can sustain trust; many officials in both sectors have already devoted years to this effort. But there are some things that the news media could do to improve their own performance:

A. Create easily accessible guidebooks for journalists on radiation science. The guidebook could be in print and online, and updated or modified in real time during a crisis. The source material for such guidebooks is already available. For example, see Armin Ansari's book, *Radiation Threats and Your Safety: A Guide to Preparation and Response for Professionals and Community* (Boca Raton, Florida: Chapman & Hall, 2010). But the material could be gathered, synthesized, and prepared in a way that today's journalists would find useful when working under tight deadlines.

B. Train journalists to better cover complex scientific topics such as radiation science and deepen the knowledge of specialized science writers. As it stands, too few journalists can master topics of physics, biology, chemistry, and other disciplines. On nuclear power, it would be worthwhile to train journalists long before an event occurs. Utilities and regulators in the United States have intensified their efforts on communications since Three Mile Island; some utilities regularly conduct drills and exercises to better prepare for an actual event. But too often, journalists are not involved, and they should be, in a transparent way that could also lead to coverage of the exercises and open discussion of risk. The Fukushima crisis showed that the most basic problem of a nuclear emergency—the problem of decay heat generated by the reactor and how to cool it—was not well understood by journalists and the public.

C. Create an independent and respected reference center to act as an easily accessible, rapid-response bridge between journalists and experts. In the nuclear field this could include access to engineers, physicists, and other scientists. A model might be the Science Media Centre in the United Kingdom, which describes itself as "a press office for science when science hits the headlines." When a major story breaks, the group provides quick access to scientists, fact sheets, and links to press officers or Web sites for more information.

D. Make widely available a publication on how to help the public and press better understand the complex technology, science, and risks in such a crisis. The International Atomic Energy Agency currently trains teams from different countries in risk communications, but does not yet have such a book, although it says one is being prepared. The global speed and power of information today demands quick access to such a publication, which should have wide distribution.[38]

E. Find ways to support, improve, and more broadly distribute the growing legion of specialized online publications and aggregators on these issues. For example, the blog armscontrolwonk.com distributed useful daily bulletins about Fukushima—but how many people knew about it? More broadly, crisis communications must be up to the speed and breadth of today's digital media.

F. Debate the kind of communications issues that would arise in the event of a serious nuclear weapons accident—one involving radiation and/or an explosion. To date, there has been almost no experience with crisis communications for such events, although nuclear weapons are under the control of national governments and militaries. The impact on public trust could be severe. Is it worth devoting time to preparations for such an emergency?

Conclusion

A broad lesson of these three crises is that in a nuclear emergency, transparency is always the best policy. Efforts to minimize danger, withhold information, or mislead people will backfire, setting off doubts that linger for decades. If civilian nuclear power is to thrive, governments, regulators, utilities, and others will need to find ways to build trust and sustain it. The challenge is not only to provide solid and timely information, but also to cope with other factors, including public fear. In communications, quality is just as important as frequency. The Japanese nuclear regulator, NISA, sent out its first release at 3:16 p.m. on March 11, and after that sent out 155 press releases and held 182 press conferences through May 31, about three times a day in the early weeks. Yet public confidence plunged as people lost faith in what they were being told.[39]

In any consideration of how to manage a nuclear crisis, journalism is a vital link between the authorities and the public. Today, journalism is in the throes of upheaval. The digital world is raw and unformed, characterized by new channels brimming with citizen reporters who can freely distribute eyewitness accounts and opinions to millions of people instantly. More traditional journalists have rapid access to huge amounts of data, yet they often lack the time or skills to synthesize or check it. As noted at the outset of this paper, news today has become a vast, transnational force—one that never stops at passport control. A failure in one country can undermine public trust in many others. Even in an authoritarian country like China, news about a disaster such as a high-speed train collision can rapidly breach the Great Firewall of censorship and be transmitted to millions of people by micro-blogging.

Training and improved understanding of science would help some journalists. But this is not enough to transform public opinion. The credibility of the nuclear enterprise rests on the actions of governments, regulators, and utilities as well. They must absorb the lessons of Three Mile Island, Chernobyl, and Fukushima: a nuclear emergency is not only a crisis of information, but of confidence. Trust is fragile, and to break it once is to lose it for a long time.

Acknowledgment

The author wishes to thank Jim Hoagland for valuable advice on this paper.

Notes

1. J. Samuel Walker, *Three Mile Island: A Nuclear Crisis in Historical Perspective* (Berkeley: University of California Press, 2004), p. 200. The author is the historian of the Nuclear Regulatory Commission and this is one in a series of commission-sponsored volumes on the history of nuclear regulation.

2. Paul Slovic, "Trust, Emotion, Sex, Politics, and Science: Surveying the Risk-Assessment Battlefield," *Risk Analysis*, vol. 19, no. 4, 1999.

3. Peter M. Sandman, "Tell It Like It Is: 7 Lessons from TMI," *IAEA Bulletin*, vol. 47, no. 2, www.iaea.org.

4. "Report of the President's Commission on the Accident at Three Mile Island," Oct. 30, 1979, p. 57.

5. See Walker, *Three Mile Island*, chapter 4, "Wednesday, March 28."

6. Ibid., p. 78.

7. Ibid., p. 80.

8. Ibid., p. 83.

9. The hydrogen spike occurred early in the accident sequence, but became a public issue only days later.

10. Walker gives this interesting account of the Hendrie remarks (*Three Mile Island*, pp. 162–170):

At 2:45 P.M. on Saturday, Hendrie met with reporters in the briefing room in Bethesda that the NRC had set up the previous day. He realized that the White House wanted [NRC spokesman Harold] Denton to be the sole source of information for the media about conditions at the plant, and had expressed his support for that arrangement in conversations with [White House press secretary] Jody Powell. But reporters covering the Three Mile Island story in the Washington area clamored for a press conference with NRC policy makers, and, at the urging of Frank Ingram of the Office of Public Affairs, Hendrie reluctantly took on the assignment on behalf of the commission. As he remembered it, Ingram told him that "somebody's got to say something . . . or they will tear the building down."

The reporters whom Hendrie addressed were not hostile but they were persistent, and he soon regretted his decision to hold the press conference. He later described it as a "disaster," largely because statements that "wouldn't have excited undue unrest" in context created a "hell of a flap" when aired in an incomplete form. Asked whether an evacuation would be necessary before dealing with the hydrogen bubble, Hendrie replied that it might prove to be a "prudent, precautionary measure." He estimated that an evacuation would extend to a distance of between ten and twenty miles in a downwind quadrant. When a reporter wondered about the chances of a hydrogen explosion, Hendrie commented that it was "a problem which is of concern and which we are working on very intensely at the moment." His own uneasiness about the bubble had somewhat diminished during the day, and he expressed the view that prevailed among the experts whom the NRC consulted, that "we are some time from any possibility of a flammable condition." Hendrie sought to be candid without causing excessive alarm, but his remarks, especially the first acknowledgement that a hydrogen explosion was a matter of concern, soon led to an eruption of panic in central Pennsylvania.

11. "Report of the President's Commission," p. 40.

12. Ibid., p. 134.

13. See David E. Hoffman, *The Dead Hand: The Untold Story of the Cold War Arms Race and Its Dangerous Legacy* (New York: Doubleday, 2009), p. 250: Vladimir Gubarev, the science editor of Pravda, who had good contacts in the nuclear establishment, heard of the accident soon after it happened and called Alexander Yakovlev, Gorbachev's close adviser and champion of new thinking. But Yakovlev told him to "forget about it, and stop meddling," Gubarev recalled. Yakovlev wanted no journalists to witness the scene. But Gubarev was persistent, and kept calling Yakovlev every day. Yakovlev finally authorized a group of journalists to go to Chernobyl, including Gubarev, who had a physics degree but also wrote plays and books. He arrived May 4 and returned May 9. His private report to Yakovlev is contained in Politburo records Fond 89, Delo 6, at the Hoover Institution Library and Archives. The private report depicted chaos and confusion. One hour after the explosion, the spread of radiation was clear, he said, but no emergency measures had been prepared. "No one knew what to do." Soldiers were sent into the danger zone without individual protective gear. They didn't have any. Nor did helicopter pilots. "In a case like this, common sense is required, not false bravery," he said. "The whole system of civil defense turned out to be entirely paralyzed. Even functioning dosimeters were not available." Gubarev said, "The sluggishness of local authorities is striking. There were no clothes, shoes, or underwear for victims. They

were waiting for instructions from Moscow." In Kiev, the lack of information caused panic. People heard reports from abroad but didn't get a single word of reassurance from the leaders of the republic. The silence created more panic in the following days when it became known that children and families of party bosses were fleeing. "A thousand people stood in line in the ticket office of the Ukraine Communist Party Central Committee," Gubarev said. "Naturally, this was perfectly well known in the city." When Gubarev returned to Moscow, he gave Yakovlev his written report. It was passed to Gorbachev.

14. Hoffman, *The Dead Hand*, pp. 244–253. For additional material, see Grigori Medvedev, *The Truth About Chernobyl* (New York: Basic Books, 1991), Evelyn Rossiter, trans.; Piers Paul Read, *Ablaze: The Story of the Heroes and Victims of Chernobyl* (New York: Random House, 1993); and Zhores Medvedev, *The Legacy of Chernobyl* (New York: W. W. Norton & Co., 1990). Also see the extensive work of the United Nations Chernobyl Forum Expert Group "Environment," including "Environmental Consequences of the Chernobyl Accident and Their Remediation: Twenty Years of Experience" (Vienna: International Atomic Energy Agency, 2006), http://www.iaea.org/NewsCenter/Focus/Chernobyl/. For a technical account of the reasons for the accident, see "INSAG-7, The Chernobyl Accident: Updating of INSAG-1," Safety Series No. 75-INSAG-7, report by the International Nuclear Safety Advisory Group (Vienna: IAEA, 1992).

15. A. N. Makukhin, First Deputy Director, Ministry of Energy and Electrification, "Urgent Report," April 26, 1986, No. 1789-2c, Dmitri Volkogonov Collection, Library of Congress, from Archive of the President of the Russian Federation, Reel 18, Container 27.

16. Michael Dobbs, *Down with Big Brother: The Fall of the Soviet Empire* (New York: Knopf, 1997), p. 160.

17. The Chernobyl Forum, 2003–2005, "Chernobyl's Legacy: Health, Environmental and Socio-economic Impacts" (Vienna: IAEA, 2006). In another estimate, at least 6,000 more died from radiation exposure, and perhaps many more. David R. Marples, "The Decade of Despair," *Bulletin of the Atomic Scientists*, May–June 1996, pp. 22–31.

18. Norimitsu Onishi, "'Safety Myth' Left Japan Ripe for Nuclear Crisis," *New York Times*, June 25, 2011, p. 1.

19. Government of Japan, "Report of Japanese Government to IAEA Ministerial Conference on Nuclear Safety—Accident at TEPCO's Fukushima Nuclear Power Stations," transmitted by the Permanent Mission of Japan to the IAEA, June 7, 2011. See http://www.iaea.org/newscenter/focus/fukushima/japan-report.

20. Ibid., p. 25. Also see IAEA, "The International Nuclear and Radiological Event Scale User's Manual," 2008 Edition, pp. 17–18.

21. Norimitsu Onishi and Martin Fackler, "In Nuclear Crisis, Crippling Mistrust," *New York Times*, June 12, 2011, p. 1.

22. Andrew Higgins, "Amid reactor crisis, head of Japanese utility vanishes," *Washington Post,* March 29, 2011, p. A1.

23. Onishi, "Safety Myth."

24. Norimitsu Onishi and Martin Fackler, "Japan Held Nuclear Data, Leaving Evacuees in Peril," *New York Times,* August 9, 2011, p. 1.

25. For a compilation of polls on nuclear questions, see after page 100 in the AEI study described here: http://www.aei.org/paper/energy-and-the-environment/polls-on-the-environment-energy-global-warming-and-nuclear-power.

26. Taiga Uranaka, "Crisis-hit Tokyo Electric posts $7.4 billion quarterly loss," Reuters, August 9, 2011. Also, on the reactor re-starts, see Associated Press, "Japan gives preliminary OK to restart 2 nuclear reactors, even with safety upgrades pending," April 9, 2012.

27. Paul Slovic, "Perception of Risk and the Future of Nuclear Power," *Physics and Society,* vol. 23, no. 1 (January 1994), a paper presented at a session on Risk and Nuclear Power at the American Physical Society meeting in Washington, D.C., April 14, 1993.

28. Slovic, "Trust, Emotion," p. 699.

29. Walker, *Three Mile Island,* p. 165.

30. "Report of the President's Commission," p. 58.

31. Ibid., p. 19.

32. Hoffman, *The Dead Hand*, p. 252.

33. Private communication with the author, August 11, 2011.

34. Onishi and Fackler, "In Nuclear Crisis."

35. Thomas Inglesby and Anita Cicero, "Early Learning from the Events in Japan," UPMC Center for Biosecurity, March 30, 2011, http://www.upmc-biosecurity.org/website/resources/publications/2011/2011-03-30_Early-Learning-from-Events-in-Japan.

36. Andrew C. Revkin. "Dot Earth" blog, *New York Times*, March 24, 2011, http://dotearth.blogs.nytimes.com/2011/03/24/warning-whiplash-with-radiation-in-the-news.

37. David Ropeik, guest blog, *Scientific American*, March 21, 2011, http://blogs.scientificamerican.com/guest-blog/2011/03/21/poor-risk-communication-in-japan-is-making-the-risk-much-worse.

38. A spokeswoman said the IAEA is nearing the final stages of publication on a document that will cover public communications in the event of a nuclear

or radiological emergency. The new publication will provide practical guidance in the form of action guides, information sheets, templates, and checklists for public information officers on the following topics: risk perception; emergency public communications; risk communication process; sources of information; audiences, communication channels, and messages; public information and communication organization; media relationships; how to be a spokesperson; and roles and coordination of information. Dana J. Sacchetti, Press and Public Information Officer, IAEA, communication to author, August 16, 2011.

39. "Report of the Japanese government," pp. IX–2.

Nuclear Enterprise Conference
October 3–4, 2011
Hoover Institution
Stanford University

<u>*Monday, October 3rd*</u>

8:00–8:30 A.M. **Continental Breakfast**

8:30–9:00 A.M. **Opening Remarks**
George Shultz & Sidney Drell

9:00–12:00 P.M. **Session I: Safety Issues—Nuclear Weapons**
Chairman: Sidney Drell
- Summaries of Prepared Papers
- Open Discussion

12:00–1:00 P.M. **Buffet Lunch**

1:00–4:00 P.M. **Session II: Nuclear Reactor Safety**
Chairman: Burton Richter
- Summaries of Prepared Papers
- Open Discussion

6:00–9:00 P.M. **Dinner**

<u>*Tuesday, October 4th*</u>

8:00–8:30 A.M. **Continental Breakfast**

8:30–11:30 A.M. **Session III: Economic and Regulatory Issues**
Chairman: George Shultz
- Summaries of Prepared Papers
- Open Discussion

11:30–12:30 P.M.	**Session IV: Media and Public Policy** *Chairman: Jim Hoagland* ▪ Summaries of Prepared Papers
12:30–1:30 P.M.	**Buffet Lunch**
1:30–3:30 P.M.	**Session IV *(continued)*** ▪ Open Discussion
3:30–3:45 P.M.	**Break**
3:45–5:00 P.M.	**Session V: Final Wrap-up** *Chairmen: George Shultz & Sidney Drell*

About the Authors

Steven P. Andreasen is a consultant to the Nuclear Threat Initiative in Washington, D.C., and teaches at the University of Minnesota's Hubert H. Humphrey School of Public Affairs. He served as director for defense policy and arms control on the National Security Council from 1993 to 2001 under President Bill Clinton and in the Department of State under presidents George H. W. Bush and Ronald Reagan. In 2002, he was the Democratic candidate for Congress in Minnesota's First District. His articles and opinion pieces have been published in many newspapers and journals.

Gary S. Becker, a senior fellow at the Hoover Institution, is a professor of economics and sociology at the University of Chicago and a professor at the Booth School of Business. He has honorary degrees from Harvard, Princeton, Columbia, Hitotsubashi University in Tokyo, University of Aix Marseille, and many other universities. He won the Nobel Prize for Economics in 1992, the Presidential Medal of Freedom in 2007, the National Medal of Science in 2000, the John Bates Clark Medal of the American Economic Association in 1967, and numerous other awards. He has written more than ten books and over 100 professional articles.

Edward Blandford is a Stanton Nuclear Security Fellow at the Center for International Security and Cooperation at Stanford University and a research assistant professor in the Chemical and Nuclear Engineering

Department at the University of New Mexico. His research interests include nuclear reactor thermal-hydraulics in support of the safety of nuclear installations, probabilistic risk assessment, physical protection strategies for critical nuclear infrastructure, and best-estimate code verification and validation. Before pursuing graduate work, he worked at the Electric Power Research Institute.

MICHAEL J. BOSKIN is a senior fellow at the Hoover Institution and the Tully M. Friedman Professor of Economics at Stanford University. He is also a research associate at the National Bureau of Economic Research and advises governments around the world on economic policy. Among other posts, he served as chairman of the President's Council of Economic Advisers from 1989 to 1993.

ROBERT J. BUDNITZ has been involved with nuclear-reactor and radioactive-waste safety for many years. He is on the scientific staff at the University of California's Lawrence Berkeley National Laboratory, where he works on nuclear power safety and security and radioactive-waste management. From 2002–07 he was at UC's Lawrence Livermore National Laboratory, during which time he spent two years in Washington developing a new science & technology program for the Department of Energy's Office of Civilian Radioactive Waste Management. He has also served on the staff of the Nuclear Regulatory Commission (1978–1980) and then ran a one-person consulting practice in Berkeley for many years.

JEREMY CARL is a research fellow at the Hoover Institution, focusing on energy and environmental policy, with particular emphasis on energy security and global fossil fuel markets. Before coming to Hoover, he was a research fellow at the Program on Energy and Sustainable Development at Stanford and a visiting fellow in resource and development economics at the Energy and Resources Institute in New Delhi, India. He is an editor of *Conversations about Energy: How the Experts See America's Energy Choices* (Stanford: Hoover Institution Press, 2010); his work has appeared in numerous books and journals in the energy and environmental fields.

John F. Cogan is a senior fellow at the Hoover Institution and a professor in the Public Policy Program at Stanford University. His current research is focused on U.S. budget and fiscal policy, social security, and health care. He has devoted a considerable part of his career to public service. He has served as deputy director at the U.S. Office of Management and Budget and as assistant secretary for policy at the U.S. Department of Labor. Mr. Cogan has also served on numerous state of California congressional and presidential advisory commissions. Additionally, he was a member of the Council of Economic Advisers for governors Pete Wilson and Arnold Schwarzenegger.

Drew DeWalt, after graduating from Notre Dame in 2004, was commissioned as an officer in the U.S. Navy and entered the nuclear submarine service. He began his career as a division officer on the USS *Cheyenne* in Pearl Harbor, Hawaii, and completed it as an assistant professor in the Naval Reserve Officers Training Corps program at Notre Dame, where he taught naval science and maritime affairs. He is a graduate student at Stanford University, studying for a joint master's degree in business administration and public policy as he pursues a career in energy policy and renewable energy resources.

Sidney D. Drell is a senior fellow at the Hoover Institution and professor emeritus of theoretical physics at the SLAC National Accelerator Laboratory, Stanford University. For many years he has advised the government on technical national security issues as a member of JASON, and on a number of advisory committees, including the President's Foreign Intelligence Advisory Board and Science Advisory Committee. His honors include the Enrico Fermi Medal presented by the Department of Energy on behalf of the president, a MacArthur Foundation Prize Fellowship, the National Intelligence Distinguished Service Medal, and election to the National Academy of Sciences.

Jim Hoagland, a 1961 University of South Carolina graduate, did postgraduate work at the University of Aix-en-Provence and Columbia University. After working at the *New York Times* International Edition, in

1966 he joined the *Washington Post*, where he won two Pulitzer Prizes and other awards as reporter, editor, and columnist. He was bureau chief in Africa, the Middle East, and Paris before becoming foreign editor and then assistant managing editor. He began a syndicated column in 1986 from Paris. Author of *South Africa: Civilizations in Conflict* (Boston: Houghton Mifflin, 1972), Hoagland is now contributing editor for the *Post*.

David E. Hoffman is a contributing editor at the *Washington Post* and *Foreign Policy* magazine. For the *Post*, he covered the Reagan and George H. W. Bush administrations, and later became a diplomatic correspondent and Jerusalem bureau chief. He also served as Moscow bureau chief, foreign editor, and assistant managing editor for foreign news. He wrote *The Oligarchs: Wealth and Power in the New Russia* (New York: PublicAffairs, 2002) and *The Dead Hand: The Untold Story of the Cold War Arms Race and Its Dangerous Legacy* (New York: Doubleday, 2009), winner of the 2010 Pulitzer Prize for general nonfiction.

Raymond Jeanloz, a professor of geophysics at the University of California-Berkeley, chairs the National Academy of Sciences' Committee on International Security and Arms Control and is a member of the government advisory group JASON. He has been an adviser to the Department of Energy and the University of California in areas related to nuclear weapons, stockpile stewardship, and international threats, about which he has written extensive technical analyses.

William F. Martin served as deputy secretary of energy, special assistant to the president, and executive secretary of the National Security Council under President Reagan. He now chairs the Department of Energy's Nuclear Energy Advisory Committee. He has served as president of the Council of the United Nations University for Peace in Costa Rica and is a co-founder of the Robinson-Martin Security Studies Program in Prague. In 2008, he was appointed by then IAEA Director General Mohamed ElBaradei as coordinator of the 20/20 Project, a study of the

IAEA's future. His publications include several books on energy statistics and forecasting.

Regis A. Matzie retired from Westinghouse Electric Company on July 1, 2009. Since then, he has provided technical consulting services to the international nuclear industry. Prior to retirement, he was senior vice president and chief technology officer at Westinghouse, responsible for all R&D and advanced nuclear plant development. Previously, he was responsible for the development, licensing, engineering, project management, and component manufacturing of new Westinghouse light water reactors. A U.S. Naval Academy graduate, he served in the U.S. nuclear submarine program for five years, followed by twenty-five years in the Naval Reserves, retiring as a captain. He received his MS and PhD degrees in nuclear engineering from Stanford University.

Michael May is a professor emeritus (research) in Stanford's School of Engineering, a senior fellow with Stanford's Freeman Spogli Institute for International Studies, former co-director of Stanford's Center for International Security and Cooperation, and director emeritus of Lawrence Livermore National Laboratory. He was a member of the U.S. delegation to the Strategic Arms Limitation Talks and is a Fellow of the American Physical Society and the American Association for the Advancement of Science. He received the Distinguished Public Service and Distinguished Civilian Service medals from the Department of Defense and the Ernest Orlando Lawrence Award from the Atomic Energy Commission.

Per F. Peterson is a professor and chair of the Department of Nuclear Engineering at the University of California-Berkeley. After working at Bechtel from 1982 to 1985 on high-level nuclear waste processing, he earned his doctorate in mechanical engineering at UC-Berkeley. He has also served as chair of the Energy and Resources Group. He was appointed by the Obama administration in February 2010 as a member of the Blue Ribbon Commission on America's Nuclear Future. He has worked on problems in energy and environmental systems, including advanced reactors, inertial fusion, high-level nuclear waste processing, and nuclear materials management and security.

Robert L. Peurifoy worked at Sandia National Laboratories for thirty-nine years in nuclear weapon RDT&E. From September 1973 to July 1983, he was responsible for weapon design at Sandia in Albuquerque. His directorate was involved in the design of five of the seven weapon types now in the stockpile. An advocate for nuclear weapon detonation safety, he introduced enhanced nuclear detonation safety technology into the stockpile and triggered the removal of nuclear weapons from quick reaction alert B-52 bombers. He retired from Sandia as a vice president in 1991 and since then has served as a consultant to several advisory groups on nuclear weapons.

Burton Richter is a senior fellow of Stanford University's Freeman Spogli Institute for International Studies, Woods Institute for the Environment, and Precourt Institute for Energy, and also Paul Pigott Professor in the Physical Sciences, emeritus. He received the Nobel Prize in physics in 1976. He is director emeritus at SLAC National Accelerator Laboratory and past president of the American Physical Society and the International Union of Pure and Applied Physics. He is an adviser to the Department of Energy on both nuclear and renewable energy. He is involved with the JASON group; the Commissariat à l'Energie Atomique; the National Research Council; industrial advisory boards; and boards of directors of several corporations.

George P. Shultz, the Thomas W. and Susan B. Ford Distinguished Fellow at the Hoover Institution, has had a distinguished career in government, in academia, and in business. He is a professor emeritus in Stanford University's Graduate School of Business and has held four different cabinet posts, most notably as secretary of state (1982–1989). He also served as secretary of labor and the Treasury, was director of the Office of Management and Budget, and was president and director of the Bechtel Group Inc.

Christopher Stubbs is a professor of physics and of astronomy at Harvard University. His research interests lie at the intersection of cosmology, particle physics, and gravitation. He also has a strong interest in national security. He is a member of JASON, a group of scientists and engineers who advise government agencies on national security issues. He also serves on the technical advisory group for the Senate Select Committee on Intelligence. He received an international baccalaureate diploma from the Tehran International School, a bachelor of science in physics from the University of Virginia, and a doctorate from the University of Washington.

John B. Taylor is a senior fellow at the Hoover Institution and a professor of economics at Stanford University. He has served on the President's Council of Economic Advisers and as undersecretary of the Treasury for international affairs. His book *Getting Off Track: How Government Actions and Interventions Caused, Prolonged, and Worsened the Financial Crisis* was one of the first on the crisis. He has since co-edited two books on preventing crises—*The Road Ahead for the Fed* and *Ending Government Bailouts As We Know Them*—and authored a third on future policy: *First Principles: Five Keys to Restoring America's Prosperity.*

Frank A. Wolak is the Holbrook Working Professor of Commodity Price Studies in the Economics Department and director of the Program on Energy and Sustainable Development at Stanford University. From 1998 to 2011, he chaired the Market Surveillance Committee (MSC) of the California Independent System Operator. In this capacity he testified numerous times at various committees of the Senate and House of Representatives on issues relating to market monitoring and market power in energy markets. He also has worked on the design and regulatory oversight of the electricity markets internationally in Europe, Australia/Asia, and Latin America, as well as throughout the United States and Canada.

Index